MICROBIAL CULTURE

ONOTT

The INTRODUCTION TO BIOTECHNIQUES series

Editors:

J.M. Graham Merseyside Innovation Centre, 131 Mount Pleasant, Liverpool L3 5TF

D. Billington School of Biomolecular Sciences, Liverpool John Moores University, Byrom Street, Liverpool L3 3AF

Series adviser:

P.M. Gilmartin Centre for Plant Biochemistry and Biotechnology, University of Leeds, Leeds LS2 9JT

CENTRIFUGATION
RADIOISOTOPES
LIGHT MICROSCOPY
ANIMAL CELL CULTURE
GEL ELECTROPHORESIS: PROTEINS
PCR
MICROBIAL CULTURE

Forthcoming titles

ANTIBODY TECHNOLOGY
GENE TECHNOLOGY
LIPID ANALYSIS
GEL ELECTROPHORESIS: NUCLEIC ACIDS
PLANT CELL CULTURE
LIGHT SPECTROSCOPY
MEMBRANE ANALYSIS

MICROBIAL CULTURE

Susan Isaac and David Jennings

Department of Genetics and Microbiology, University of Liverpool,
P.O. Box 147, Liverpool L69 3BX, UK

βIOS
SCIENTIFIC
PUBLISHERS

© **BIOS Scientific Publishers Limited, 1995**

First published 1995

A CIP catalogue record for this book is available from the British Library.

ISBN 1 872748 92 9   l000 Sc8 2S9

BIOS Scientific Publishers Ltd
9 Newtec Place, Magdalen Road, Oxford OX4 1RE, UK
Tel. +44 (0) 1865 726286. Fax +44 (0) 1865 246823

DISTRIBUTORS

Australia and New Zealand
    DA Information Services
    648 Whitehorse Road, Mitcham
    Victoria 3132

Singapore and South East Asia
    Toppan Company (S) PTE Ltd
    38 Liu Fang Road, Jurong
    Singapore 2262

India
    Viva Books Private Limited
    4346/4C Ansari Road
    New Delhi 110002

USA and Canada
    Books International Inc.
    PO Box 605, Herndon, VA 22070

Typeset by Chandos Electronic Publishing, Stanton Harcourt, UK.
Printed by Information Press Ltd, Oxford, UK.

Contents

Abbreviations

A	adenine
ACDP	Advisory Committee on Dangerous Pathogens
C	cytosine
COSHH	Control of Substances Hazardous to Health
DAPI	4', 6-diamidino-2-phenylindole
DNA	deoxyribonucleic acid
EDTA	ethylenediaminetetraacetic acid
G	guanine
HPLC	high performance liquid chromatography
RFLP	restriction fragment length polymorphism
RNA	ribonucleic acid
RNase	ribonuclease
rRNA	ribosomal RNA
SEM	scanning electron microscopy
T	thymine
TEM	transmission electron microscopy
UV	ultraviolet
v/v	volume/volume
w/v	weight/volume

Preface

Microbial Culture is intended to be an introductory text for those who are about to start growing micro-organisms. Such persons may be undergraduates, graduates, technicians or postdoctoral scientists. We anticipate that they are about to start their work in a laboratory in which some microbiological work is already being undertaken. So the emphasis here is on work at the bench with the necessary back-up of equipment and laboratory facilities. It could be said that if this is the case there should be no need for a book of this kind. But in our experience, the converse is true. In most microbiological laboratories, the knowledge possessed by a person joining is often assumed. Indeed, that knowledge may be tested not by prior questioning but by mistakes which can prove at a minimum inconvenient, but possibly very costly in terms of time, money and safety. In any case, passing on information just by word of mouth is never very efficient. On all counts, there is a need for a text to give persons new to growing micro-organisms the necessary background.

With the above in mind, we have written this book based on our own experiences, both good and bad! The great diversity of micro-organisms has meant that we have eschewed, for the most part, providing detailed protocols. We have taken the approach of highlighting matters to keep in mind when carrying out microbiological work. Effective protocols give every detail of the procedures to be followed. Circumstances, such as the organism used, the facilities available, etc., can all conspire to make it difficult to follow a protocol as written. Then there must be adaptation, and the success of such depends upon a proper understanding of what is involved. We hope this book will help with that understanding and provide a mental check-list of precautions and the like, which, if taken into account, will lead to the successful adaptation of any protocol to the specific needs of the investigator.

Throughout the text and with a supplementary list in Appendix C we have provided guidance about further reading. While that reading can provide the necessary assistance to address the particular problem of an investigator, we do recommend making contact with an expert in the field if it is proposed to start upon a field of microbiology totally new to the investigator's laboratory. It is in similar spirit that occasionally we stress the need to make contact with the appropriate local expert over a particular issue, about which it is difficult to deal with all aspects within the somewhat limited confines of the present text.

Finally, we realize that often microalgae and protozoa are considered to be micro-organisms. We have made passing reference to these microbes but the bulk of the text concerns bacteria and fungi. This is partly due to our own experience, but also because we believe persons working with algae and protozoa are better served, because of the nature of the organisms, by referring to the relevant specialized texts on these organisms.

We are most grateful to our colleague Dr Alan McCarthy for reading the manuscript and giving us much helpful comment and advice. Chapter 6 has benefitted very considerably from the input of Dr Ron Burke who read the first draft. We thank him for his help, as indeed we are pleased to thank our editor, Dr John Graham, for his patience and constructive criticism.

Susan Isaac
David Jennings

1 Introduction

1.1 Why culture microbes?

In order to be able to identify the kinds of microbes present in a sample, to estimate the profusion in which they are present and to investigate their particular metabolic properties one has to be able to cultivate them. In addition, by using laboratory cultures the number of variables, both chemical and physical, imposed on the organisms at any one time can be limited and controlled, allowing their responses to external factors to be assessed.

Cultivation aids accurate identification. It is necessary to grow up sufficient cells (biomass) of a species in order to be able to examine it and to establish its growth requirements. In the first instance, use of culture media which relate to the natural conditions in which the organism grows will usually allow cultivation, with subsequent experimentation leading to more precise definition of specific physiological requirements and genetic traits.

Microbes are extremely versatile and can show considerable morphological and/or physiological variation. They are to be found in almost every ecological niche, even seemingly hostile situations (such as hot springs), where it would be reasonable to assume that no organism could have the capacity to survive, let alone grow and reproduce. Additionally, many microbes readily form associations with other microbes, higher plants and animals as symbionts, parasites and pathogens. Such associations can range from those with mutually beneficial arrangements to those which have harmful or detrimental effects on one or both partners. Indeed, some microbes are not able to survive, grow and reproduce unless they remain in association with other organisms.

Large numbers (hundreds of thousands) of microbial species have been defined, each with its own particular characteristics and requirements. More are being discovered and described, some with very unusual or unlikely life styles. It would be difficult therefore, if not impossible, to have a clear understanding of all microbial species and how each species is defined. Certainly, it is unnecessary to have such a detailed knowledge to be able to manipulate and successfully cultivate a wide range of micro-organisms. Here we provide guidelines to the principles and aims of microbial cultivation. In similar vein, in later chapters only pointers to more specialized methods are given, since these are obviously outside the scope of any single text.

Experimental work often has several aims. These might include the need to learn something about a very wide range of microbial species, inhabiting a particular ecological niche. Alternatively, very broad knowledge of a single species, or an in-depth study of one specific biochemical pathway operated by a species, may be required. Whatever the eventual aim or outcome of an investigation, it is clear that methods for growing individual species in *sterile* laboratory culture will be needed, and for this cells are usually placed into prepared growth medium. Such cultures are often termed *axenic,* meaning pure, uncontaminated cultures of a single organism. The researcher therefore requires an understanding of a range of techniques in order to become familiar with the organism(s) in which he or she is interested:

- pure culture for the isolation of single cells from the natural environment to an appropriate sterile growth medium;
- aseptic handling for successful transfers between different kinds of growth media;
- composition and preparation of growth media for different purposes;
- safe handling and storage of the isolated organisms in order to protect both the cultures and all higher organisms.

These topics will be covered in the first part of this volume, more specialized methods following in the second part.

As a group, micro-organisms have a very wide range of distinguishing characteristics, some of which can be determined simply and rapidly, while others require more time and specialized approaches:

- morphology of cells and colonies, for example shape, staining properties, motility, differentiated structures;
- growth characteristics, for example colony formation, nutrients needed, physical conditions;
- biochemical and molecular properties, for example secondary products formed, surface antigens, molecular characteristics (GC ratios, RNA sequences).

In general, microbes are very small and, as a result of a high surface area to volume ratio, are able to take up nutrients and lose toxic metabolites quickly. This often results in a very high metabolic rate. Generation times of 20–30 min have been recorded under conducive conditions. As a result, it is generally possible to obtain sufficient biomass of microbes in culture relatively quickly, which is a great aid to experimental investigations.

1.2 An introduction to the classification of micro-organisms

Most micro-organisms can only be seen properly with the aid of a microscope. However, many produce colonies that can be seen with the naked eye, and some form large, visually recognizable structures, for example mushrooms and algal fronds. There are huge numbers of different species of microbes, and classification groups together those with similar life styles and morphological and physiological characteristics. Some basic information about microbial classification is given here, indicating the major differences between the groups. The term micro-organism is usually used loosely, to cover (*Table 1.1*) viruses, bacteria (including actinomycetes and cyanobacteria), algae, fungi (including yeasts, lichens and slime molds) and protozoa.

TABLE 1.1: *A working classification of micro-organisms*

Organisms	Characteristics
Viruses	Noncellular organization
Prokaryotes	No nuclear membrane, no cell organelles
Eubacteria	Small, unicellular, often rods or cocci
actinomycetes	Filamentous organisms often found in soils
cyanobacteria	Photosynthetic prokaryotes
Archaebacteria	Small, unicellular
Eukaryotes	Nucleus delimited by membrane
Algae	Majority contain chlorophyll and photosynthesize like plants, others heterotrophic. May be unicellular and microscopic or multicellular and up to several meters in length
Fungi	Rigid cell walls, unicellular or filamentous and multicellular, frequently invasive of living and nonliving substrates. Heterotrophs, lack chlorophyll
yeasts	Unicellular, reproduce by fission or budding
slime molds	Lack rigid cell wall, form amoeboid aggregations
lichens	Symbiotic association between fungus and usually a single species of alga or cyanobacterium, resistant to harsh environmental conditions
Protozoa	Single celled, ingest particulate nutrients. Often flagellate, ciliate or amoeboid

The viruses fall uniquely into a group of their own since their organization is very different from that of cellular organisms. They are not capable of independent growth and are usually handled with particular specialized techniques. Viruses will not be considered further in this volume.

The ordering of species based on evolutionary relationships is known as phylogeny. Following the recent development of molecular sequencing techniques it has become possible to make comparisons between macromolecules in living organisms. This approach (molecular phylogeny) has shown that there are three cell types and has resulted in the definition of three kingdoms (now referred to as domains), each with very distinct evolutionary differences: Bacteria (formerly Eubacteria), Archaea (formerly Archaebacteria) and Eukarya (eukaryotes, including fungi, animals, plants and protista). Micro-organisms are also often divided on the basis of whether they are prokaryotes or eukaryotes. The distinguishing features are various (*Table 1.2*) but perhaps the most easily observed is that of size; prokaryotes are usually much smaller than eukaryotes. The arrangement of nuclear material is also a major feature. Prokaryotes (Bacteria and Archaea) do not have a nuclear membrane; the genetic material is carried in a single chromosome lying inside the cell. Eukaryotes (Eukarya) on the other hand, have nuclei; the genetic material is contained within a nuclear membrane in the cytoplasm and is packaged into a number of linear bodies or chromosomes. Other discrete structures (organelles), associated with particular biochemical activities, are also contained within the cytoplasm of eukaryote cells, accounting for their larger size.

1.2.1 Prokaryotes

Archaea (Archaebacteria). The cell walls of members of the Archaea consist of proteins or unique polysaccharides but those of bacteria contain peptidoglycan. The cell membranes also differ in structure. The membranes of archaebacteria are constructed from lipids bonded to glycerol by ether linkages, whereas in membranes of eubacteria straight chain fatty acids are ester linked to glycerol. There are also differences in the metabolism of the two divisions; some archaebacteria show highly specialized metabolic sequences, for example under anaerobic conditions some (methanogens) produce methane from carbon dioxide and hydrogen and a few can convert acetate to methane.

Archaebacteria are often found in unusual habitats. Some archaebacteria are extremely salt tolerant (extreme halophiles) and

TABLE 1.2: *Major differences between prokaryotes and eukaryotes*

Prokaryotes	Eukaryotes
Usually small	Larger than prokaryotes
Genetic material not surrounded by membrane	Nucleus membrane-bound
One chromosome	More than one linear chromosome
No meiosis	Mitosis and meiosis
No mitochondria	Most have mitochondria
Oxidative metabolism in cell membrane	Oxidative metabolism in mitochondria
No endoplasmic reticulum or Golgi apparatus	Endoplasmic reticulum and Golgi apparatus
Ribosomes (70S) dispersed in cytoplasm	Ribosomes (80S) attached to endoplasmic reticulum
No chloroplasts	Chloroplasts site of photosynthesis (if photosynthetic)
Photosynthesis in cell membrane (if photosynthetic)	Photosynthesis in chloroplasts
Peptidoglycans present in wall	No peptidoglycan
Flagella with one fibril	Flagella with 9+2 arrangement of microtubules and membrane

are found in salt lakes. They have a very high requirement for salt and are capable of growth only at near NaCl saturation. Others grow only at extremely high temperatures (100–110°C) and are dependent on elemental sulfur for growth. These are found in hot sulfur-rich springs, volcanic habitats and geothermal heated areas (vents) under the sea. Some (methanogens) produce methane under anoxic and anaerobic conditions. Methanogens are found in habitats rich in organic matter where oxygen levels are low, for example lake muds, swamps and the rumen of herbivores. Carbon dioxide, methyl-containing compounds or acetate are converted to methane by this group.

Bacteria (Eubacteria). The bacteria are subdivided primarily on the presence or absence of a cell wall. Those lacking walls are the mycoplasmas. The remaining bacteria are classified into groups (*Table 1.3*) based on the chemical composition of the cell wall (Gram stain; see Section 8.3.2), morphology, nutritional requirements and metabolic activity [1]. General accounts of the major groups of

TABLE 1.3: *Working classification of Bacteria and Archaea*

Group I   Gram-negative bacteria with cell walls
Autotrophs
  Photoautotrophs (CO_2 fixed using light energy)
    Oxygenic, photosynthetic bacteria (cyanobacteria), e.g. *Microcystis, Anabaena, Nostoc*
    Anaerobic, anoxygenic photosynthetic bacteria, e.g. *Chromatium, Thiospirillum, Rhodobacter*
  Chemoautotrophs (CO_2 fixed by oxidation of inorganic compounds)
    Ammonia and nitrite oxidizers, e.g. *Nitrosomonas, Nitrobacter*
    Sulfides and sulfur oxidizers, e.g. *Thiobacillus, Beggiatoa*

Heterotrophs
  Aerobes/microaerophiles, rods and cocci, e.g. *Agrobacterium, Pseudomonas*
    Pathogens of man and animals, e.g. *Brucella, Neisseria, Bordatella*
  Facultative anaerobes, motile or nonmotile rods, often in association with plants and animals,
    Enterobacteriaceae-oxidase negative, e.g. *Erwinia, Escherichia, Salmonella*
    Vibrionaceae-oxidase positive, e.g. *Aeromonas, Vibrio*
  Obligate anaerobes, isolated from intestinal tracts of man and animals,
    sewage sludge and anoxic muds, e.g. *Bacteroides, Veillionella*
    Includes sulfate reducers, e.g. *Desulphobacter*

Miscellaneous
  Spirochaetes, e.g. *Leptospira, Treponema*
  Curved bacteria, e.g. *Campylobacter*
  Rickettsias and chlamydias (intracellular obligate parasites)
  Budding and appendaged bacteria, e.g. *Caulobacter*
  Sheathed bacteria, e.g. *Leptothrix*
  Gliding bacteria, e.g. *Cytophaga*
  Myxobacteria, fruiting bacteria, e.g. *Myxococcus*

Group II   Gram-positive bacteria with cell walls
All are chemo-organotrophs (heterotrophs)

Endospore formers
  Form resistant endospores, often motile, aerobic or facultative aerobes, e.g. *Bacillus;* obligate anaerobes, e.g. *Clostridium*

Lactic acid bacteria
  Do not form spores, rods or cocci, catalase negative, microaerophiles, e.g. *Lactobacillus, Lactococcus, Streptococcus*

Actinomycetes and relatives
  Most form branching filaments, mostly aerobic or facultative anaerobes, e.g. *Mycobacterium, Streptomyces, Frankia*

Gram-positive cocci
  Chemo-organotrophs, do not form endospores, catalase positive,
    facultative anaerobes, e.g. *Staphylococcus*
    obligate aerobes, e.g. *Micrococcus*
    obligate anaerobes, e.g. *Peptococcus*

Group III   Bacteria lacking cell walls
Mycoplasmas
  Facultatively anaerobic to obligately anaerobic, some motile (gliding), e.g. *Mycoplasma, Spiroplasma*

Group IV   Archaea
Methanogens
> Strict anaerobes, form methane as main metabolic end-product, e.g. *Methanococcus, Methanobacterium*

Archaeal sulfate reducers
> Strict anaerobes able to form H_2S from sulfate, traces of methane in addition. Extremely thermophilic (up to 92°C), e.g. *Archaeoglobus*

Extremely halophilic archaebacteria (halobacteria)
> Aerobic or facultatively anaerobic chemo-organotrophs. Require sodium chloride in concentrations approaching saturation, e.g. *Halobacterium, Halococcus*

Archaebacteria lacking cell walls
> Thermoacidophilic, aerobic, coccoid, e.g. *Thermoplasma*

Extremely thermophilic archaebacteria
> Obligate thermophiles, optimal growth 70–105°C, metabolize sulfur, e.g. *Thermoproteus*

---

bacteria (*Table 1.3*) can be found in a number of microbiology texts. However, a number of points deserve special mention here.

Often bacteria live as single cells but some daughter cells do not separate after division and as a result groups, or chains of cells are formed which tend to stay together in the environment, forming a *colony*. On the surface of a suitable medium bacterial colonies appear as rounded, shiny dots (resembling blobs of nail varnish) which are usually pale in colour, ranging from buff to pink, yellow or white, and may be opaque or translucent. Some colonies may have characteristic margins but usually the appearance is unremarkable. Cellular morphology can be determined relatively easily. Bacterial cells are usually spherical (coccus, cocci), rod-shaped (bacillus), curved (vibrio) or spiral (helix). Other information about the cells in a culture will involve the use of staining techniques and various diagnostic tests. Some bacteria produce highly resistant spores within their cells (*endospores*). These have a great capacity to withstand much higher temperatures than vegetative cells and are also resistant to desiccation and chemical disinfectants. The position of endospores within bacterial cells is used in conjunction with other characteristics in classifying some bacterial species.

The actinomycetes are a large group of filamentous, Gram-positive (see Section 8.3.2) bacteria. These species have the dimensions of bacteria (0.5–1.0 μm diameter) but grow in a filamentous, ramifying form which is very like the mycelial growth more usually associated with fungi. The actinomycetes are divided into eight groups according to morphological, biochemical and metabolic characters. They include some species of extreme importance; for example, the streptomycetes includes 500 species of *Streptomyces*, many of which produce secondary metabolites, such as antibiotics, with enormous value in medicine and veterinary science. Many species produce spores with characteristic structures which are used in identification. These spores impart a dusty appearance to the surface of colonies which can lead to their confusion with sporulating fungal cultures.

The cyanobacteria are a large group of oxygenic phototrophic species, formerly known as the blue-green algae. There is no doubt about their prokaryotic status, however, and the term cyanobacteria should now be used. These diverse organisms may be unicellular or filamentous and vary enormously in size (1–60 µm). Cyanobacteria have photosynthetic pigments (chlorophyll *a*) and accessory pigments (phycobilins). Some species form heterocysts, which are sites of nitrogen fixation, at intervals along filaments. The enzyme responsible for nitrogen fixation (nitrogenase) is very oxygen sensitive and it is likely that the thick heterocyst walls limit the diffusion of oxygen into these cells, thus acting as a protective mechanism for the enzyme system.

Photosynthetic purple and green bacteria also use light as an energy source but these are anoxygenic phototrophs, that is, no oxygen is produced during photosynthesis. They use bacteriochlorophyll (of which there are several forms) and one photosystem for photosynthesis. In the presence of oxygen, bacteriochlorophyll synthesis is inhibited and therefore photosynthesis cannot proceed. They are usually found in the anaerobic regions of lake muds and marshes where light intensity is low.

Some bacteria make use of inorganic energy sources and have a vital role to play in the cycling of elements in the environment, both on a localized and a global scale. Sulfur-reducing bacteria use sulfate (SO_4^{2-}) as terminal electron acceptor, forming elemental sulfur which is then converted to hydrogen sulfide (H_2S) or thiosulfate ($S_2O_3^{2-}$). These bacteria are common in anaerobic conditions in lakes and pools, giving rise to the pungent (bad eggs) smell of anaerobic muds as levels of H_2S accumulate. Some sulfur-oxidizing bacteria (e.g. *Thiobacillus*) live in aerobic conditions and oxidize sulfide (S^{2-}) to sulfate (SO_4^{2-}), whereas others, the anaerobic sulfur oxidizers (purple or green sulfur bacteria) require low light intensities to function in anaerobic conditions where they oxidize sulfide (H_2S) and elemental sulfur to sulfate. These organisms are important in sulfur cycling in the environment. Iron-oxidizing bacteria oxidize reduced ferrous iron (Fe^{2+}) to ferric iron (Fe^{3+}) and $Fe(OH)_3$ accumulates around the cells, often in gelatinous sheaths.

Some bacteria (nitrifying bacteria) derive energy by oxidizing ammonia (NH_3) to nitrite NO_2^- (ammonia-oxidizing bacteria) and then other species oxidize NO_2^- to nitrate (NO_3^-) (nitrite-oxidizing bacteria) using carbon dioxide as a source of carbon. Other bacteria carry out *denitrification*, reducing nitrate to nitrogen gas under anaerobic conditions, a process by which nitrogen is lost from the substrate. Under normal circumstances nitrogen is constantly lost from soil by leaching with water and incorporation into living organic matter. Nitrogen is also added to soil by bacteria by fixation into ammonia and these organisms therefore have a very important role in nitrogen cycling in the environment. The species concerned are free-living aerobic bacteria (e.g. *Azotobacter, Beijerinckia*) as well as members of the cyanobacteria. Some nitrogen fixers live in association with other organisms and these fix much larger amounts of nitrogen and often improve the nitrogen status of growing plants. Species of *Rhizobium*, living in close association with the roots of leguminous plants, are the most active in nitrogen fixation.

1.2.2 Eukaryotes

Fungi. Fungi are a very versatile and diverse group of eukaryotes (*Table 1.4*). They are highly invasive and heterotrophic, frequently secreting enzymes allowing penetration of otherwise insoluble substrates and making potential nutrients accessible which can then be absorbed and utilized. They often occur as single-celled forms, or with poorly differentiated thalli. More usually, however, vegetative growth is as *filamentous hyphae* (singular, *hypha*) extending through the substrate, and as the result of tip growth and repeated branching a colony, or *mycelium,* is formed. Fungal mycelium is often visible to the naked eye. Mycelium often grows over the surface of the substrate but hyphae also penetrate into it. Additionally, aerial hyphae are formed which often bear asexual spores above the surface of the colony. These spores are often pigmented (most commonly green, brown, white, yellow) and give the surface of the colony a dry, dusty appearance, characteristic of the fungus. Spores are most usually formed on mature hyphae at the central region of colonies. Some species of fungi produce pigments which are released from hyphae into the external medium.

Sexual spores are usually produced in smaller numbers but are larger and more resistant and can withstand adverse environmental conditions. These often have thick outer coverings (ornamented and pigmented) which are formed closer to the substrate, often shrouded by vegetative mycelium. Sexual spores may be formed within structures which act as additional protection. Some sexual spores remain viable after years of apparent desiccation. Sexual spores are often liberated into the atmosphere for dispersal by air currents and the fungi use many different mechanisms to achieve this effectively. Mushrooms and toadstools, for example, are very obvious structures on which sexual spores are formed and from which they are released for dispersal.

Many species exhibit a variety of nuclear phases during their life cycle. Vegetative hyphae may contain more than one nucleus (*coenocytic*) and even in instances where cross walls (*septa*) divide a hypha into compartments more than one nucleus (sometimes many) may occur per compartment. Additionally, nuclei and cytoplasm may move through pores in septa. Some aged areas of fungal mycelium may become devoid of cytoplasmic contents during growth; the breakdown products of cytoplasmic constituents are passed to the physiologically active regions of the colony.

Fungi are placed into taxonomic groups primarily by morphological characteristics, particularly those relating to the differentiation of spores; the way in which spores are formed and the structures that hold or enclose spores being the most important criteria. Fungal groups are also divided on the basis of cell wall components (*Table 1.4*). The slime molds (Myxomycota) do not have true cell walls but occur as free-living protoplasts, aggregating at times during the life cycle. The true fungi (Eumycota) are grouped into five subdivisions. The names of the subdivisions end with the suffix '-mycotina' (as in Basidiomycotina) but the names of members of the subdivisions end with the suffix '-mycete' (as in basidiomycete).

Within the Eumycota the Mastigomycotina and Zygomycotina are usually regarded as the lower fungi and the Ascomycotina, Basidiomycotina and Deuteromycotina as higher fungi. The latter groups have septate mycelia and are considered to be more advanced in evolutionary terms. The Deuteromycotina (Fungi Imperfecti) includes about 17 000 species grouped together through the lack of a sexual state (perfect state or *teleomorph*), most are *anamorphs* (imperfect or asexual state). The other subdivisions of the Eumycota are grouped by the characteristics of the teleomorph. It is often regarded that the Deuteromycotina are grouped together as a matter of convenience. Many of the fungi share common features with members of the Ascomycotina or Basidiomycotina. For some species a sexual stage is undescribed or may have been lost. In some cases some of the functions of the sexual stage have been replaced by the *parasexual cycle* (recombination taking place within the mitotic cycle). Many common species, some with economic significance, are placed in the Deuteromycotina, for example *Penicillium* and *Fusarium,* some

TABLE 1.4: *The major taxonomic subdivisions of fungi (from ref. 2)*

Taxonomic group	Vegetative phase	Septa	Major cell wall components	Spores	
				Asexual	Sexual
Mycota (slime molds)					
Myxomycota	Plasmodium	None	None	Zoospores	Zoospore fusion
Eumycota (true fungi)					
Mastigomycotina	Filamentous, diploid	Aseptate	Cellulose, glucan	Zoospores	Oospores
Zygomycotina	Filamentous, haploid	Aseptate	Chitin, chitosan	Sporangiospores	Zygospores
Ascomycotina	Filamentous, haploid	Septate	Chitin, glucan	Conidia	Ascospores
Basidiomycotina	Filamentous, diploid	Dolipore septum, clamp connections	Chitin, glucan		Basidiospores
Deuteromycotina	Filamentous, haploid	Septate	Chitin glucan	Conidia	No sexual stage known

are predacious on nematodes and some are important plant pathogens.

A broadly acceptable classification scheme for fungi was published by Ainsworth [3] and was adopted by Webster [4] for his useful volume covering many aspects of fungal taxonomy (see also ref. 5).

Yeasts are essentially those higher fungi which are unicellular and most are classified as Ascomycetes, although some important yeast species are Basidiomycetes. Most of their growth characteristics are akin to those of the fungal group to which they belong. Many yeast species are industrially important, whereas others are important human and animal pathogens, and for this reason yeast physiology has been relatively well documented. Certainly baker's/brewer's yeast (*Saccharomyces cerevisiae*) is one of the most studied eukaryotic organisms. It is interesting to note that some species of fungi have the capacity to grow as yeast-form cells under some conditions, but also as filamentous forms in different environments (*dimorphism*). A number of dimorphic fungi are medically important and therefore have received a great deal of attention.

Lichens (or lichenized fungi) are often classified separately. They are extremely long-lived organisms and are able to withstand very harsh environmental conditions, but are nevertheless very sensitive to atmospheric pollution. The vegetative part of a lichen is known as the thallus and is composed of a fungal partner (*mycobiont*) and a green alga or cyanobacterium (*phycobiont*). The characteristics of the lichen thallus are quite different from any of those of either partner growing individually. Over 13 500 species of fungi are known to form lichen associations (mostly Ascomycetes and a few Basidiomycetes). Many of these species are not found free-living but appear to depend on the association for survivial. Only relatively few algal species are involved in lichen symbiosis (mainly green algae) and just 12 genera of cyanobacteria.

Algae. The algae are a large group of eukaryotes. Most contain chlorophyll and photosynthesize, evolving oxygen. They are divided taxonomically by the specific pigments they contain, by the mechanism of motility and by the composition of their cell walls (*Table 1.5*). They may be small and unicellular, or, if cells aggregate rather than separating, they are said to be colonial. In some cases cells remain attached longitudinally and these are termed filamentous. There are algae with very large cells (strictly coenocytes because they contain many nuclei). These cells, which can be seen easily with the naked eye, have been used extensively for physiological investigations since electrodes and other probes can be

TABLE 1.5: *Characteristics of major groups of algae*

Group	Morphology	Pigments composition	Major cell wall components	Reserve substances	Motility
Chlorophyta (green algae)	Unicellular to multicellular	Chlorophyll *a, b,* carotenoids	Cellulose	Starch	Equal flagellae
Phaeophyta (brown algae)	Filamentous to multicellular and macroscopic	Chlorophyll *a, b*	Cellulose	Laminarin motile	Zoospores
Chrysophyta (golden-brown algae, diatoms)	Unicellular	Chlorophyll *a, c,* carotenoids	Silica	Lipids	Apical flagellae
Rhodophyta (red algae)	Multicellular	Chlorophyll *a, d,* phycocyanin	Cellulose	Starch	Nonmotile
Pyrrophyta (dinoflagellates)	Unicellular	Chlorophyll *a, c*	Cellulose	Starch	Two lateral flagellae
Euglenophyta (euglenoids)	Unicellular	Chlorophyll *a, b*	Absent	Paramlyon	Apical flagellae

inserted readily. Some algae can be very large, such as seaweeds, which may reach many meters in length.

The cell walls of algae are usually composed primarily of cellulose but with other polysaccharides, pectin, xylans and mannans included. In some cases calcium carbonate deposits are also laid down in the walls, which provide strength and rigidity. Many algal cells are motile by means of flagellae. Some species have simple polar flagellae while others have one, two or more of differing position and length. Many species are nonmotile but may have motile gametes, formed only during sexual reproduction. The diatoms move by gliding over solid surfaces.

Protozoa. Protozoa are unicellular eukaryotes which generally move to find food materials and are classified by their mechanisms of motility (*Table 1.6*). They are found in aquatic habitats or in very damp regions. Some protozoa are animal parasites and can cause debilitating, serious diseases of man and other animals (e.g. sleeping sickness, giardiasis, amoebic dysentery, malaria). They usually lack cell walls and ingest soluble food materials by invagination of the cell membrane into a food vesicle (*pinocytosis*), or take in particulate material by engulfment (*phagocytosis*). Some species are found in the

TABLE 1.6: *Characteristics of the major groups of protozoa*

Group	Motility	Growth habits	Representatives
Mastigophora (flagellates)	Flagellate	Aquatic, animal parasites	*Trypanosoma* *Giardia*
Sarcodina (amoebas)	Amoeboid	Freshwater and marine animal parasites	*Amoeba*
Ciliophora (ciliates)	Ciliate	Animal parasites, rumen inhabitants	*Paramecium*
Sporozoa (sporozoans)	Nonmotile	Animal/insect parasites, vectors	*Plasmodium* *Toxoplasma*

guts of animals, both large and small. Indeed, protozoa have an important role in the rumen of herbivores.

1.3 Nutritional modes and the exploitation of external resources

Microbes are sometimes grouped together according to different nutritional modes, although these groups are not mutually exclusive. Some organisms have the ability to adopt different nutritional modes under different conditions.

Saprotrophy describes the use of nonliving organic materials for growth. Obligate saprotrophs do not have the capacity for any other mode of nutrition. Many saprotrophic species can be grown in culture and in chemically defined media but not all saprotrophs have been isolated into culture. Some species (particularly some fungal species) are very competitive and aggressive. These may excrete compounds, and lytic enzymes, that inhibit the growth of neighboring microbes. Other species are not at all aggressive and may quickly succumb to antagonism.

Necrotrophy occurs when an organism invades and kills the living tissues of another organism, subsequently using those tissues saprotrophically. Obligate necrotrophs often have very low competitive ability.

Biotrophs invade the living cells of an organism (host) to obtain nutrient resources. This implies that any defence mechanisms of the host have been overcome and usually the host is not seriously disabled until the biotroph is near to reproduction. Obligate biotrophs have no capacity in nature to be free-living but grow and reproduce solely on the living tissues of the host. As a result it is not surprising that many biotrophs as yet cannot be maintained in laboratory culture and many have never been isolated into culture.

In the natural environment microbes grow in close proximity to a range of other organisms. In some instances growth and development may be enhanced by interactions and relationships between organisms, and in other cases there may be detrimental effects.

- *Mutualistic relationships* – both organisms are deemed to derive benefit from the association between the partners, for example lichens, mycorrhizal fungi, *Rhizobium* and leguminous plant root nodules.
- *Commensal relationships* – one organism derives benefit from a relationship while the other is unaffected, for example commensal protozoa from animal guts, leaf-surface fungi.
- *Pathogenic/parasitic relationships* – one organism benefits from contact to the detriment of the other, for example rust fungi and powdery mildews cause disease in plant hosts, bacterial infections in humans and animals.

1.4 Initial steps in dealing with microbes

In the normal course of experimental work it would be unusual to have to deal with all groups of micro-organisms at once. It might be necessary to identify all the organisms occurring in a particular sample, but in that case it would be unlikely that very detailed classification would be required. It is always possible to call on the advice of an expert for a precise identification to be made, or to have it checked. This is most advisable if an isolate from the natural environment is to be the basis for a research program. It is crucial that the organism that you plan to use for experiments has been correctly identified so that you can acquaint yourself with any possible hazards in using it, and if the work carried out is to be published.

It is, however, important and useful to realize how quickly and easily some decisions can be made to narrow down the possibilities. Most of the isolates obtained from a natural sample will be fairly common and will fall easily into particular categories. With the use of specific laboratory media and a microscope it is possible to make many valuable observations on a culture without further sophisticated equipment or training. For example, it is relatively easy to tell whether a particular isolate is bacterial, fungal or algal, and subsequent steps can be taken to make more specific identification using techniques and media relevant to that isolate.

Following visual examination of a sample (see Section 8.2) it is necessary to grow the organisms of interest in culture. Remember, however, that what an organism does in culture may not fully reflect its activity in the natural environment. In culture microbes receive easily used nutrients in large quantities and do not have to compete with other organisms or cope with fluctuating conditions. Placing microbes from a natural environment into culture may immediately tend to favor one particular microbe from a sample. It may be necessary to carry out selective or enrichment cultures to investigate fully all the components of an ecosystem and to ensure that slow-growing or unusual isolates are not overlooked. It is often difficult (or impossible using some methods) to decide which organisms appearing in culture are active in the natural environment and which are present but dormant (see Chapter 7).

Additionally, it is important to consider the possibility that organisms apparently present in a sample are contaminants, reaching the culture from the laboratory environment rather than from the sample itself. Microbes occur everywhere and the ideal conditions devised for a particular species are often ideal for many others too, which are all quick to exploit an opportunity. Spotting contamination becomes easier with experience. Contamination is often a frustrating aspect of microbiology and may occur when least expected and usually at the most awkward stage of the work. It can be overcome by practice and a little understanding of the procedures involved.

References

1. Holt, J.G., Kreig, N.R., Sneath, P.H.A., Staley, J.T. and Williams, S.T. (eds) (1994) *Bergey's Manual of Determinative Bacteriology* (9th edn). Williams and Wilkins, Baltimore, MD.
2. Isaac, S. (1992) *Fungal–Plant Interactions*. Chapman & Hall, London.
3. Ainsworth, G.C. (1973) in *The Fungi: An Advanced Treatise* (G.C. Ainsworth, F.K. Sparrow and A.S. Sussman, eds). Academic Press, New York, Vol. IV B, pp. 1–7.
4. Webster, J. (1980) *Introduction to Fungi* (2nd edn). Cambridge University Press, Cambridge.
5. Hawksworth, D.L., Sutton, B.C. and Airsworth, G.C. (eds) (1983) *Ainsworth and Bisby's Dictionary of the Fungi* (7th edn). Commonwealth Mycological Institute, Kew, Surrey.

2 Safety

2.1 General

Microbes can incapacitate and kill other organisms, including humans, so safe procedures are required for handling them. Unless such procedures are used there are risks to human health. More positively, the use of safe procedures ensures effective experimentation and the production of data of high integrity. Poor procedures militate against such data and at the same time increase the reservoir of possible contamination in the laboratory. Thus poor procedures have a psychological and financial cost through untrustworthy results. In contrast, safe procedures bring rewards over and above their requirement to prevent microbes being a hazard.

Safety is the responsibility of every scientific investigator. In whatever country you may be working, at a minimum, you must be aware of the legislation of that country covering safe working in the laboratory and the safe handling of microbes. For instance, in the UK, safe working in laboratories is covered by the requirements of the Health and Safety Act of 1974. Thus every investigator needs to know what procedures are safe for the particular work he or she is undertaking. Before starting any laboratory procedure, you need to be aware of, have assessed and recorded, any possible hazards both in the procedures themselves and the microbes that might be used (in the UK, this assessment will be over and above Control of Substances Hazardous to Health (COSHH) regulations). Your assessment will be aided by good experimental design and the determination of the requirements for glassware, chemicals and equipment. Naturally, basic to any assessment of possible hazards is knowledge about the microbe being studied. If one is using pure cultures, it is not difficult to establish the possible hazards. On the other hand, when attempting to isolate particular microbes from a substrate containing other (possibly unknown) micro-organisms, the hazards will be much less clearly defined. Whatever the situation, it is the microbes being studied which govern the procedures and facilities to be used. Since micro-organisms can be hazardous, we describe such procedures and facilities as *containment*. There are three elements of containment: laboratory practice and technique, safety equipment and facility design (*Table 2.1*).

TABLE 2.1: *Summary of recommended biosafety levels for infecting agents (reproduced from ref. 1 with permission from Cambridge University Press)*

Biosafety level	Practices and techniques	Safety equipment	Facilities
1	Standard microbiological practices	None: primary containment provided by adherence to standard laboratory practices during open-bench operations	Basic
2	Level 1 practices plus: laboratory coats; decontamination of all infectious wastes; limited access; protective gloves and biohazard warning signs as indicated	Partial containment equipment (i.e. class I or class II biological safety cabinets) used to conduct mechanical and manipulative procedures that have high aerosol potential that may increase the risk of exposure to personnel	Basic
3	Level 2 practices plus: special laboratory clothing; controlled access	Partial containment equipment used for all manipulations of infectious material	Containment
4	Level 3 practices plus: entrance through changing room; shower on exit; all wastes decontaminated on exit from facility	Maximum containment equipment (i.e. class III biological safety cabinet or partial containment equipment in combination with full-body, air-supplied, positive-pressure personnel suit) used for all procedures and activities	Maximum containment

2.2 Organisms

Micro-organisms can be classified according to the severity of the hazard. In the UK, the requisite guidance is given in a report

TABLE 2.2:   *Classification of micro-organisms according to biological hazard and their shipping requirements (reproduced from ref. 1 with permission from Cambridge University Press)*

Class I	Agents of no recognized hazard under ordinary conditions
Examples	*Saccharomyces cerevisiae, Trichoderma reesei, Lactobacillus casei*
Shipping	Culture-tube in fiberboard or other container. Permits as required
Class II	Agents of ordinary potential hazard
Examples	*Aspergillus fumigatus, Candida albicans, Cryptococcus neoformans, Staphylococcus aureus*
Shipping	Culture-tube wrapped in absorbent material, placed in metal screw-cap can, placed in fiberboard container. Permits as required
Class III	Pathogens involving special hazard
Examples	*Coccidioides immitis, Ajellomyces capsulatum, Bacillus anthracis, Yersinia pestis*
Shipment	Culture-tube heat-sealed in plastic, wrapped in absorbent material, placed in hermetically sealed can, placed in sturdy cardboard box. Permits as required. Etiologic agent warning label necessary
Class IV	Pathogens of extreme hazard
Examples	*Arthroderma simii, Pasteurella multocida,* certain animal/plant viruses
Shipment	Culture-tube heat-sealed in plastic, wrapped in absorbent material, placed in hermetically sealed can, placed in sturdy cardboard box. Required permits. Etiologic agent warning label necessary

produced by the Advisory Committee on Dangerous Pathogens (ACDP) *Categorisation of Pathogens According to Hazard and Categories of Containment* [2].

Table 2.2 shows the manner by which micro-organisms can be categorized into four classes, according to their hazardous nature and their shipping requirements, giving the names of a few representative micro-organisms and indicating the laboratory facilities required for handling them. If you are starting microbiological laboratory work, you must realize that you would need to undertake specific training before handling micro-organisms which are in classes III and IV. For that reason, we shall not deal in this book with the details of handling these organisms. Nevertheless, there is a need to be aware of the existence of organisms in these particular two classes when isolating microbes from natural substrates. It is possible that some class III micro-organisms might be isolated.

There are some points to be made about fungi. First, while most filamentous fungi are not known to have adverse effects on general human health, some of these fungi produce copious quantities of spores, which can be either asexual or sexual, and which are readily released into the air. These spores can give rise to allergic (hypersensitive) reactions in some individual workers. In particular, asthmatic breathing can be triggered which may be severe. Consequently all practices should be designed to minimize the release of spores. In this respect, laminar flow cabinets – those that blow a current of initially sterile air over the inoculation and other handling operations (see Section 2.3) – can be a hazard to the operator; under such circumstances it is better to inoculate in still air. Fungal genera whose spores are known to give rise to allergic responses are *Aspergillus, Botrytis, Cladosporium, Helminthosporium, Mucor, Penicillium, Rhizopus, Serpula* and *Ustilago*. Secondly, fungi can produce highly toxic compounds in their media, thus while the organism may not be a hazard, the spent culture may be. Thirdly, while most filamentous fungi are not harmful to man, many can be plant pathogens. So you must keep this in mind if you are handling filamentous fungi and working alongside or in the same building as investigators working with plants.

At all times you must remember that many micro-organisms not normally associated with human disease can be opportunistic pathogens causing infection of the young, the aged and in immuno-deficient or immunosuppressed individuals. Also, vaccine strains which have undergone multiple *in vivo* passage should not be considered avirulent because they are called 'vaccine strains'. Finally, always make sure that you have no exposed cuts or scratches; they should be properly covered with an effective medical dressing.

2.3 Laboratory facilities and decontamination

The term basic laboratory, used in *Table 2.1*, is a space in which experimental work is carried out with viable micro-organisms not associated with disease in healthy adults. Such a laboratory can be used with infectious agents when the hazard levels are low and laboratory workers can be protected by standard laboratory practice. Work, for the most part, can be conducted on the open laboratory bench, though certain operations have to be carried in specially designed containment cabinets. Such a cabinet (class II) is required for those manipulations which have the potential to produce aerosols which increase the exposure risk to the laboratory worker. These cabinets are essentially boxes, open to the operator directly or via gloves, which have a lower (negative) pressure inside than outside

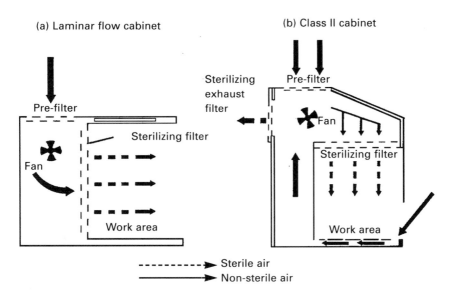

(a) Laminar flow cabinet

(b) Class II cabinet

FIGURE 2.1: *Air flow diagrams for (a) an horizontal laminar flow cabinet and (b) a class II recirculating cabinet. Reproduced from ref. 3.*

such that air is drawn into the cabinet (*Figure. 2.1b*). The outlet to a cabinet of this kind must be located such that the air expelled cannot contaminate other parts of the laboratory or the rest of the building. The cabinet just described must not be confused with those cabinets for inoculating nonhazardous organisms into cultures. These are called *sterile hoods* or *laminar flow cabinets* and should be considered as clean benches over which is blown filtered air (freed of microbial propagules) (*Figure 2.1a*). In these there is a positive pressure generated which militates against contamination from the laboratory.

As far as you yourself are concerned, some particularly important points are listed as follows.

- Always wear a laboratory coat, which buttons up effectively, while working in the laboratory. The coat should not be taken out of the laboratory except for laundering, when the coat should first be autoclaved.
- Any pipetting should be by mechanical means. Mouth pipettes must not be used. Likewise, eating, drinking, smoking and applying cosmetics are not permitted in the laboratory.
- Always keep your bench-top clean such that it is a poor source of microbes. Use a solution of the detergent issued in your laboratory for cleaning glassware.

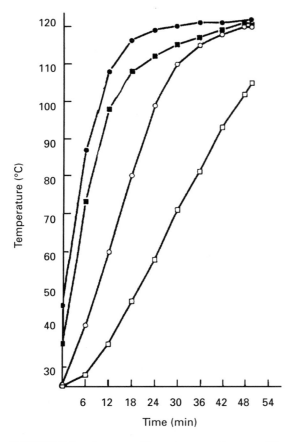

FIGURE 2.2: *Average time–temperature profiles of waste loads (1.75 kg) of Petri dishes (100 x 15 mm) containing agar. The loads were autoclaved in a steel cylinder with (●) or without (○) 1 liter of water, and in a steel container in an autoclavable plastic waste bag with (■) or without (□) 1 liter of water. The autoclave reached 121–122°C within 3 min after the cycle was initiated. Reproduced from ref. 4 with permission from the American Society for Microbiology.*

Microbiological waste (such as cultures on agar) should be autoclaved, even when it might seem that the waste does not present a significant hazard. Unwanted old cultures frequently pick up contaminants which might be more dangerous to health than the original culture. Autoclaving is carried out on the waste contained in autoclave-proof plastic bags. The processing time required to achieve necessary decontamination will depend on specific loading factors. As one can see from *Figure 2.2* the time taken to reach 115°C can vary very considerably depending on the container used for the waste and the presence of water vapor. It may therefore be necessary to have a processing time of 60–90 min to decontaminate waste. Once decontaminated, microbiological waste can be treated as refuse, providing that the bags are not opened after autoclaving.

If dry-heat ovens are used for decontaminating glassware, the time taken will be much longer, because dry heat is much less effective than wet heat in destroying microbes. Usually glassware, of all kinds, is submerged for a period of time in disinfectant – a solution of a chemical that will kill those micro-organisms coming into contact with it but not having an untoward effect on humans if their skin is exposed to it over a short period (nevertheless, when handling disinfectant one should always wear gloves). The classic disinfectant is carbolic acid, the use of which in antiseptic surgery was pioneered by Lord Lister; but now there are many different disinfectants. You are advised to use that disinfectant being used in the laboratory in which you are working. One disinfectant which can be recommended is Stericol™ (Lever Industrial) at 1–2% (v/v) concentration. Stericol™, which is phenolic-based, also has detergent properties, making it a particularly useful agent in the processing of contaminated glassware.

Surfaces in the work area of containment cabinets should be wiped down with a clean disposable cloth containing 70–80% (v/v) ethanol solution. Prior to this the fan should have been on for 5–10 min. When all culturing operations are completed the surfaces should again be wiped down with ethanol solution. Cloths for wiping down should be treated as microbiological waste. The effectiveness of a cabinet will depend on the filter; they should be cleaned frequently following the manufacturer's instructions. It is possible that the cabinet will need more drastic treatment to get rid of contaminating organisms. This is best carried out using formaldehyde gas. But you must consult your safety officer before carrying out such a procedure. A crucial aspect of the operation is the location of the exhaust from the cabinet. At the very simplest, one discharging into the laboratory must be exhausted at another location. But if the exhaust is into a duct, there could be problems due to uncertainties as to the route followed by the duct. During the operation the cabinet must be sealed completely from the laboratory.

2.4 The key piece of advice about safety

Most accidents are caused by carelessness, through the failure to think about the safety implications of a chosen procedure. Often the lack of sufficient thought is due to misplaced confidence, based on ignorance. *If you are at all uncertain about an appropriate procedure, then ask for advice*. Certainly, never assume that you know how to use unfamiliar equipment. The need to be prepared to seek advice will be a recurring theme throughout this book.

References

1. Hawksworth, D.L. and Allner, K. (1988) in *Living Resources for Biotechnology: Filamentous Fungi* (D.L. Hawksworth and B.L. Kirsop, eds). Cambridge University Press, Cambridge, pp. 54–75.
2. Advisory Committee on Dangerous Pathogens (1984) *Categorisation of Pathogens According to Hazard and Categories of Containment*. Her Majesty's Stationary Office, London.
3. Morgan, S.J. and Darling, D.C. (1993) *Animal Cell Culture*. BIOS Scientific Publishers, Oxford.
4. Vesley, D. and Laner, J. (1986) in *Laboratory Safety: Principles and Practices* (B.M. Miller, ed.). American Society for Microbiology, Washington, DC, pp. 182–98.

3 Media Preparation and Components

3.1 Introduction

Micro-organisms are extremely versatile and also very variable in their modes of nutrition. Chemical elements are taken in from the environment and incorporated into the structure of microbial cells by the metabolic processes of the particular organism. Some only require a few inorganic materials for growth whereas others need complex organic compounds. Nutrients are therefore required in different amounts in culture depending on the organism concerned and its capabilities. Laboratory culture systems are sometimes designed to simulate natural conditions, providing those nutrients likely to be found in the usual habitat in order to increase the chances of isolating specific microbes. However, it is more usual that a nonspecific medium is used to encourage very rapid accumulation of biomass or the liberation of extracellular enzymes or secondary products. In many instances, quantities of cells are grown and used for chemical analyses and physiological measurements with little prior investigation into the state of those cells or consideration of the stage of development.

When an organism is first placed into culture it takes time to adapt to those conditions. Proliferation is slow at this stage (*lag phase*) but, in a conducive culture, there will be a gradual increase in metabolic activity and the growth rate will increase rapidly (*acceleration phase*) until the rate reaches a maximum for those particular conditions (*exponential phase*). Such rapid growth cannot be sustained for a long period in a batch culture because key nutrients will be depleted rapidly, or toxic products may build up and the growth rate will be limited (*deceleration phase*). As nutrient levels become limiting, as is particularly the case with actinomycetes and fungi, cells may switch from *primary metabolism* and production of biomass to *secondary metabolism*, often producing novel products which are not apparently needed for growth but which may be important for survival in the natural environment. Such products are formed as a result of switches in biochemical pathways, although they may confer some form of competitive advantage in nature. A reduction in the supply of key nutrients results in the accumulation of some biochemical intermediates, and when this occurs secondary pathways are initiated. Some secondary products are highly toxic to other organisms, for example mycotoxins, antibiotics.

Once an organism has been isolated into culture, it is then possible to place limitations on that culture (e.g. omitting particular nutrients or altering the formulations) in order to determine specific requirements and understand the patterns of metabolism for that particular isolate. It is important to remember that not all microbes have yet been grown in laboratory culture.

3.2 Formulation of laboratory culture media

A huge number of recipes exist for culture media designed for the growth of micro-organisms [1]. Some are highly specific, for the more fastidious microbes, and some are very general. This chapter will consider the principles and approaches to cultivation. It must, however, be remembered that for some organisms it is essential to use specialist methods and published procedures should be consulted.

Some nutrients are required in relatively large amounts while others are essential but are needed in smaller quantities. The depletion of one (or more) form of nutrient from a culture may limit the growth of the organism, and it is very important, particularly for physiological experiments, that the effects of *nutrient balance* are understood. When an organism is first placed into culture it is usual to supply nutrients in excess.

In order to grow and develop, microbes require an energy source, a supply of carbon and other appropriate elements in both a usable form and at an appropriate concentration. Micro-organisms are often grouped together by the mode of metabolism and the source of energy they use (*Table 3.1*). *Autotrophs* have relatively simple requirements, which are chemically inorganic. Such microbes need carbon dioxide as the source of carbon and they derive energy from either light (radiant

TABLE 3.1: *Nutritional strategies*

	Energy source	**Carbon source**
Chemoautotrophs	Inorganic compounds	Carbon dioxide
Chemoheterotrophs	Organic compounds	Organic compounds
Photoautotrophs	Light	Carbon dioxide
Photoheterotrophs	Light	Organic compounds

energy) or the oxidation of inorganic compounds. *Heterotrophs* use organic compounds as energy sources and derive energy from organic compounds or from light.

Carbon supply. Although carbon dioxide can be used as a major source of carbon by species of autotrophs, most nonalgal microbes are heterotrophs and require organic compounds to provide carbon and a supply of energy. A very wide range of materials can be utilized (particularly amino acids, peptides, organic acids, monosaccharides, oligosaccharides, polysaccharides, acetate, fatty acids and hydrocarbons), and these provide the main sources of carbon skeletons from which cells can synthesize other compounds.

Most heterotrophic micro-organisms require large amounts of carbon (usually supplied as glucose in culture). Levels are quickly reduced in a culture such that the production of biomass is limited. Different carbon sources will influence the final biomass produced by an organism in culture and this requires investigation if high yields are required. When organisms are provided with a mixture of carbon compounds, there is usually preferential use of one form first, even though the organism may have the capacity to use all the forms, and may well do so in prolonged culture. In some instances, the provision of a mixture of carbon sources may enhance microbial growth quite markedly, giving rise to a greater biomass than would be produced on the individual sources alone.

Supply of other elements. The requirements of microbes for elements other than carbon is essentially as one would expect from a knowledge of other organisms. Most microbial biochemistry texts will indicate probable roles. The elements that are required for microbial functioning can be classified in terms of the concentrations at which they will be present in an effective growth medium. Thus, there are the major elements, which are (along with carbon) nitrogen, sulfur, phosphorus and potassium, required at around 10^{-2} M and higher. Then there are the macroelements, such as magnesium, calcium, iron, copper, manganese and zinc, required at 10^{-5}–10^{-3} M. Finally, there are the microelements (sometimes called trace elements), which are required at very low concentration. Molybdenum, for instance, need only be present at 10^{-9} M, though it should be noted that this particular element is only required by those microbes utilizing nitrate. Other trace elements are needed by some organisms but are not essential overall. Thus, cobalt is needed by those microbes that synthesize their own vitamin B_{12}, while nickel is necessary for those exhibiting urease activity.

When considering nitrogen, sulfur and, to a much lesser extent, phosphorus, one needs to keep in mind that these elements can be supplied in both inorganic and organic form. Sometimes only the latter can be utilized, in which case the compounds are not only a source of the particular element but they are also a source of carbon. Also, a wide variety of organic compounds containing nitrogen, sulfur or phosphorus can be accessible to microbes. In similar manner to the situation when a microbe is growing in a medium containing two or more sugars, when there is sequential use of the individual sugars, there is preferential utilization of certain nitrogen, sulfur and phosphorous compounds over others which may also be present in the medium. There is now a great deal of information for particular microbes about molecular processes that control the uptake and metabolism of nitrogen compounds, so that there is use of one nitrogen compound rather than another. Essentially the requisite transport system(s) and metabolic pathways for a particular nitrogen compound are fully operational while the equivalent process for other nitrogen compounds are nonfunctional.

The requirement for an element by a microbe can change according to the stage of growth, particularly if there is onset of secondary metabolism or there is differentiation. These events take place frequently when the nitrogen content of the medium is much reduced in the presence of very much greater concentrations of carbon. Reduction of other nutrients in similar manner can lead to similar changes, though, in the case of secondary metabolism, the nature of the metabolites produced will differ. The message from this is that the culture medium for effective secondary metabolism and significant differentiation will be necessarily different from that used to promote the production of biomass.

The minor elements are metals, and when present in the form of cations are often in nature either tightly bound on to solid surfaces or are in insoluble form, as is the case with iron and manganese. This is true particularly of iron. In consequence, the amount of free metal ion in solution can be very low. Where the matter has been investigated, it has been found that microbes have very high-affinity specific systems for the uptake of minor elements. Furthermore, many bacteria and fungi are known to produce compounds (siderophores) which can scavenge iron from the external medium such that the metal iron becomes more readily available to the cell. There is evidence that other types of compound are produced for some of the other minor elements but the matter needs further investigation.

In terms of growing microbes in culture, we can highlight a number of matters to keep in mind with respect to the above observations. These are listed as follows.

- Although it is likely that for maximum growth the minor elements are required at significant concentration, in actual fact routine growth of many microbes can be achieved at very much lower concentrations because of their ability to scavenge the element from the medium.
- As a result of this ability to scavenge, in most cases minor and trace elements in a medium can come as what might be termed 'impurities' in other reagents used, or as components of the extract forming the basis of the medium. The main exception to this is magnesium, which is nearly always required at near millimolar concentration.
- It follows from what has just been said that, because minor and trace elements can be scavenged by micro-organisms, it can be very difficult to show the requirement for such an element, because of the technical problems involved in producing media totally free of the element under consideration. It is for this reason that there is still argument about whether or not fungi need calcium for growth. More convincing evidence for a requirement for an element, about which there is doubt as to whether or not it is required by a microbe, is more likely to come from biochemical studies.
- Iron can often become unavailable to a microbe in a culture medium. To obviate this possibility, a simple chelating agent, such as citrate or ethylenediaminetetraacetic acid (EDTA), is added to the medium. From what we know from studies on higher plants it is likely that the iron is taken into the cell while still associated with the chelate.

Finally, there is a need to keep in mind the fact that metals such as copper and zinc are toxic at concentrations higher than those normally used in growth media. The concentrations that are toxic will depend on the microbe, since some microbes have evolved a considerable degree of tolerance to such metals and also to other metals which are similarly toxic but are not required for the growth of the microbe.

Growth factors. There is some difference of opinion as to what consitutes a microbial growth factor. There have been two definitions:

- a growth factor is a compound which is required in small amounts for growth, excluding those compounds that function as structural material, and is not used as an energy source;
- a growth factor is a compound which, in minute amounts, is necessary or stimulatory for growth but does not serve as an energy source.

The first definition covers true vitamins, those compounds having a catalytic function in the organism as a coenzyme or a constituent part of a coenzyme (*Table 3.2*). The second definition clearly encompasses a broader range of compounds than the first and covers compounds such as amino acids, nucleotides, fatty acids and sterols. It should be noted that not only will such compounds be incorporated into structural material within the microbial cell but these compounds are required at a much higher concentration than the vitamins (10–1000 pg cm^{-3} in the former case vs. 0.01–1 pg cm^{-3} in the latter). In spite of the fact that the second definition covers essentially two classes of compound, it is probably the more satisfactory as an operational definition at the present time.

When a microbe cannot be grown in defined culture it is likely that a nonvitamin growth factor is required. Identification of that factor is extremely difficult because, if the obvious candidates have been tried and found not to produce growth, finding the appropriate compounds is then virtually a matter of chance. A microbe that is unable to synthesize a growth factor is said to be *auxotrophic* for that particular compound.

Since growth substances are required at relatively low concentrations, conditions favoring breakdown of components of a culture

TABLE 3.2:   *The role of vitamins in microbial metabolism*

Vitamin	Function
p-Aminobenzoic acid	Precursor of folic acid biosynthesis
Folic acid	Cofactor in synthesis of methionine, thymine and pantothenic acid
Biotin	Fatty acid biosynthesis, CO_2 fixation
Pantothenic acid	Component of coenzyme A
Nicotinic acid	Oxidation–reduction reactions
Thiamin	Decarboxylation of α-keto acid
Riboflavin	Oxidation–reduction reactions
Vitamin B$_6$	Amino acid transformations
Vitamin B$_{12}$	Synthesis of deoxyribose and methionine
Lipoic acid	Pyruvate and α-ketoglutarate oxidation

medium will have a much more dramatic effect on growth factors than on major components (in terms of concentration) of a culture medium. For this reason, individual growth factors should be filter sterilized and added to the remainder of the medium which has been sterilized by autoclaving (see Section 4.1). There is no need for filter sterilization if growth factors are added in bulk in the form of, say, yeast extract.

3.3 Determination of nutrient requirements

In isolating an unknown microbe from the environment it is usual to place it first on to a complex or *undefined medium*, which is generally based on natural materials, for example blood, rumen fluid or, more generally, extracts thereof, for example liver, malt, soil, yeast extracts. Some undefined media have a somewhat bizarre basis, such as the now famous corn steep liquor used in the early production of penicillin.

Subsequently, once the nutritional requirements are better understood the organism may be placed on to either a *semi-defined* or a *defined medium*. The latter medium is one in which the precise chemical composition is essentially known, any ignorance can only concern trace elements whose concentration is not easily determined, because either the necessary analytical techniques are not available or the extent of their being contaminants of other nutrients cannot be assessed. A semi-defined medium is usually one in which the majority, or all, of the major nutrients are defined but the trace elements and any growth factors are added in bulk, such as yeast extract.

Some general comments about the best use of the three types of media are given as follows.

Undefined media. Sometimes called complex media for the obvious reason that their composition may only be known in broad terms and, more importantly, there are almost certainly more than the necessary minimum number of components present to support growth. It is preferable when choosing an undefined medium to choose one that mimics the environment in which the organism is found, or which improves upon it in terms of nutrient availability. As indicated above, there is an enormous range of undefined media. Inspection of a handbook for culturing a particular group of microbes will both confirm this statement and demonstrate the wide variety of such media. The great virtues of undefined media are that, if well chosen, they support good growth of the relevant organism, with no loss of vigor over many subcultures and they are both relatively easy and relatively inexpensive to make up. The economic aspect is important, particularly if large amounts of media are being used. Undefined media are almost always used for routine culturing purposes.

Semi-defined media. Once growth has been achieved in culture and the broad physical requirements have been investigated and met, for example temperature and light, then it is possible, in a series of experiments, to replace the main undefined components with known compounds, making the medium semi-defined. When growing microbial biomass for experimental purposes, frequent use is made of semi-defined media in which the major nutrients are defined but the trace elements and growth factors remain undefined, being added to the medium in the form of small amounts of yeast extract, casein hydrolyzate or bacteriological peptone. In this way, biomass with consistent properties can be produced in a repeatable manner using a growth medium that is relatively easy and cheap to produce.

Defined media. It will be clear from the foregoing that defined media, because the nature and concentration of all the components of the medium need to be known, can be costly in terms of time, effort and money to produce. That is assuming that the medium has been devised to have all the components at an optimum concentration (i.e. that the medium supports the highest possible growth rate). If one has to produce such a medium *ab initio,* much experimentation is required to determine the optimum concentration for all components. That said, defined media are often required for precise biochemical and physiological studies. If such studies are clearly required, then the use of a defined medium is unquestioned; if such studies are not planned, some thought is needed before embarking on the use of a defined medium. It should be noted that defined media have yet to be devised for many microbial species. Indeed, in some cases *in vitro* culture is not yet possible, even on complex media.

Nutrient requirements can be determined through a program such as the following; each stage may require a series of experiments.

- A culture is initiated from a natural source on to a suitable undefined medium and subsequently subcultured for storage of the organism.
- The organism is subcultured on to a semi-defined medium in which the main nutrient sources are varied in concentration, singly, from a high level (above normal) to zero.
- Components that are required for growth and those that stimulate growth can then be determined.

- Complex carbon source(s) can be replaced with a range of defined sources individually and at various concentrations.
- Complex nitrogen sources can be replaced by alternative sources individually.
- Complex nitrogen sources can be replaced by an amino acid mixture (such as casamino acids) and, if growth occurs, concentrations of amino acids can be varied singly and in combination.
- If a source of growth factors, such as yeast extract, is used, this can be replaced by a solution of vitamins. If growth occurs, the concentrations of vitamins can be varied singly and in combination. Minor and trace elements will have to be added separately.

Liquid/solid media. Both liquid (*broth* when undefined) and solid (or semisolid) media are used routinely for culturing microbes, and each has advantages.

Solid cultures can be inspected easily for microbial growth, colony diameters can be assessed, pigment production can be monitored, etc. In general, the gelling agent used is *agar.* Agar becomes molten above 82°C and remains liquid as it cools, until about 40°C when it solidifies. It is ideal for microbiological techniques because it can be mixed with nutrients in solution before it solidifies. Additionally, after it has solidified, agar medium can be incubated at virtually all potential growth temperatures (−10 to 80°C), including those used for most thermophiles. Agar is a complex polysaccharide derived from seaweeds. It is a natural extract and contains variable amounts of vitamins at trace concentrations and ions, particularly potassium and sodium, but, in general, it is not used by microbes to support growth. A range of grades, or qualities, of agar can be obtained commercially. In general, the more purified the agar, the less extraneous nutrients are present and the higher the cost. In some instances a medium may be solidified with gelatin, but this is used as a major nutrient source by a range of microbes and, additionally, it becomes molten at 23°C. Silica gel is the most appropriate alternative to agar but is less easy to use and contains a high concentration of monovalent cations.

Liquid cultures are best for the measurement of microbial biomass and study of metabolic products produced in the medium. After incubation, biomass can easily be separated from the medium and washed, so the composition of both the organisms and the residual culture medium can be determined (see Chapter 6).

Agitated (shake)/static cultures. When liquid broth cultures are inoculated and incubated undisturbed, then only the surface layers will remain aerobic. The lower regions quickly become oxygen depleted and the system becomes progressively more heterogeneous.

It is more usual for liquid cultures to be incubated with agitation (Chapter 6). The agitation (either by shaking or stirring) results in better mixing of nutrients and better aeration of the medium. Very high growth rates can be achieved in nutritionally well-balanced, stirred, liquid cultures, and this principle is used in fermentation technology.

The production of some fungal metabolites can be carried out in static liquid cultures to considerable effect. When a plug of mycelium is placed on to the surface of a liquid culture and incubated without agitation, hyphae will grow out from the inoculum in the same way as on solidified medium. A floating raft of mycelium develops, with aerial and submerged hyphae. The colony may sporulate in the same way as on a solid medium. After growth, the culture fluid may be drained away from beneath the mycelium and the biomass may be collected for dry weight determination.

3.4 Selective media

Some microbes are very fastidious and will only grow on very specific media in which all their growth requirements are met. Others are more versatile and will grow freely on various media. When making first isolations from the natural environment, the numbers of microbes likely to be cultivated from a single small sample are very large. In this situation it is possible to use selective media that will tend to favor, or to suppress, a particular group of microbes. Growth, even on a suitable and rich medium, will also be influenced by physical conditions [2].

Use of specific physical regimes can result in the encouragement or suppression of particular microbes. For example, an acidulated medium will favor organisms with low pH growth optima. Lowering the pH of a culture medium is often used as a strategy to favor fungal growth. Temperature regimes can be used to similar effect, encouraging the growth of organisms with either high or low growth optima. In addition, it is possible to add *antimicrobial compounds* to culture media in order to suppress the growth of particular microbes. The purpose of the experiment must be carefully considered, to ensure that any added compound is not likely to affect the metabolism of the organism under test. Streptomycin (50–100 mg l^{-1}), gentamycin (50 mg l^{-1}) and cycloheximide (50 mg l^{-1}) interfere with protein synthesis in bacteria. One of these antibacterials can be added to a culture medium to restrict the growth of Gram-positive and Gram-negative bacteria (see Section 8.3.2). Nystatin ($10–200$ U ml^{-1}) affects the permeability of cell membranes and is used to restrict the growth of fungi in culture.

Specific *diagnostic tests* have been developed for bacteria to distinguish between those organisms with different physiological properties in terms of their response to specific growth conditions. The metabolic capabilities and nutrient requirements of bacteria are important factors in the classification of these organisms. Consideration of all selective media is beyond the scope of this text, but key tests are covered in Chapter 8 and handbooks of tests for specific bacterial groups can be consulted [2].

3.5 Nonculturable microbes

Where organisms cannot be isolated into laboratory culture, other means must be devised for studying them. In the case of pathogens, particularly plant pathogens, it is possible to maintain an isolate by inoculation of a susceptible host. The life cycle of the pathogen will then be completed in the host tissues in the usual way and spores can be transferred subsequently to healthy plants. In the case of pathogens that will also grow in culture, it is important to realize that the capacity for pathogenicity and host plant infection may be lost in the subculturing system. A passage through a host plant at intervals will help to maintain the pathogenicity of the isolate.

Obligate symbionts can often only be isolated and identified by means of their spores and cannot be grown in culture (e.g. mycorrhizal fungi, lichenized fungi). Isolations must always be made from the natural environment.

References

1. Atlas, R.M. and Parks, L.C. (eds) (1993) *Handbook of Microbiological Media*. CRC Press, Boca Raton, FL.
2. Gerhardt, P., Murray, R.G.E., Costilow, R.N., Nester, E.W., Wood, W.A., Kreig, N.R. and Phillips, G.B. (eds) (1981) *Manual of Methods for General Bacteriology*. American Society for Microbiology, Washington, DC.

4 Principles and Initial Steps in Culturing

4.1 Aseptic techniques and manipulations

The handling and culture of micro-organisms are carried out using aseptic techniques, the primary aims of which are to keep the test microbe in, and other microbes out of, the culture. This ensures that the microbial culture under test, the laboratory worker and the surroundings are kept free from contamination. At first, some of the operations may seem unwieldly and difficult to carry out but, as with other specialist techniques, practice helps to improve success rate.

All apparatus and culture media must be sterilized prior to use, so that it is completely free from microbial contamination. Any equipment that comes into contact with microbes must also be sterilized after use and before disposal. It is most usual to sterilize laboratory equipment by heating with steam, under pressure (autoclaving) although a conventional pressure cooker can be used for small-scale sterilization. By heating at increased pressure, the temperature is raised above 100°C and therefore the time for which heating is required can be kept as short as possible. Autoclaving can affect the components of nutrient medium but the effect can be minimized (but *not* removed) by limiting the heating time. Normal autoclave conditions are 15 min at 121°C (15 lb in^{-2}; 10^2 kPa). Temperature and pressure can be closely controlled. The sterilization chamber of the autoclave may be aligned vertically, most usual for large-capacity versions, or horizontally. During sterilization, solutions may boil vigorously and bubble within vessels. It is, therefore, advisable to place laboratory media and solutions for autoclaving in large containers, for example autoclave a 100 ml solution in a vessel with a 200–250 ml capacity. For autoclaving, screw caps of vessels should be placed on loosely and tightened after vessels have been removed from the autoclave, after sterilization. Heat-sensitive tape can be attached to items prior to sterilization. After heating, the tape develops dark stripes and therefore it is easy to check which items have been processed. It is advisable to seek expert advice locally the first time you use an autoclave.

Autoclaving is a very efficient sterilization system since microbial cells are more susceptible to killing by heat when wet and more resistant when dry. However, it is sometimes necessary for particular items of equipment to be kept dry (usually empty glassware, pipettes, glass spreaders). In that case it is necessary to sterilize these items by exposure to high temperatures (160–180°C) for longer times (2 h) for effective sterilization.

Some media components can be degraded or react with other components during autoclaving and, when this is so, the components must be sterilized separately and added later using aseptic techniques. Precipitation of material in the culture medium is an obvious sign that autoclaving is having an effect; color change is another. Those components that are not heat stable (e.g. vitamins) must therefore be sterilized by filtration with *bacteriological filters* which remove microbial cells from solutions. Such filters have very small pores (0.2–0.45 µm) and retain microbial cells on the surface, so that sterile solution is collected after filtration. *Membrane filters* are composed of cellulose acetate which acts as a sieve; *nucleopore filters* are manufactured from polycarbonate films and have very precise pore sizes, retaining any particles larger than the pores. All equipment used for filtration must be sterilized and kept sterile prior to use. It is necessary, therefore, to wrap small filters in aluminum foil or to place them inside larger containers for autoclaving. Additionally, the tops of flasks and bungs (see Section 4.2.6) need to be covered in this way so that the rims do not become contaminated prior to use. Prior to autoclaving, pipettes are normally plugged with a small bung of cotton wool, at the end which will be handled. This is done so that, when the pipette is in use, air passing to and from the barrel, via the pipette filler, travels through sterilized cotton wool. Pipettes are then placed inside a metal container with the delivery tip innermost, or individually wrapped in aluminum foil with the tip end indicated. Sterilized pipettes must be unwrapped at the end to be handled so that the delivery tip is kept sterile. In many laboratories today disposable pipettes are purchased sterilized and individually sealed ready for use. In addition to using sterile equipment, it is important that all glassware is clean so that vessels can be poured from easily without dripping, and without leaving drops behind.

Before culturing micro-organisms, choose a clear space for working in and clean it carefully with disinfectant (see Section 2.3) before starting. Keep it clear and orderly as work progresses. Before you start an experiment, think carefully which steps are needed to ensure that the micro-organisms under test are kept *in* the culture vessels designed for them and to keep potential contaminants *out*. Following a few guidelines will help to ensure success in culturing micro-organisms.

- Make sure that disinfectant is near to hand in case of spillage; treat any spill *immediately*.
- Work is always carried out in close proximity to a lit Bunsen burner (this is required for flaming tops of tubes and other implements, to sterilize them).
- Vessels which are involved in the culturing procedure are opened for only the minimum amount of time, just as they are needed.
- As the top is removed from a vessel the neck is quickly passed through the Bunsen flame to kill any microbes at the opening, and similarly, the neck is passed through the flame again after making transfers in to or out of the vessel.
- *Never* place the tops of vessels on the bench, they are held in the hand and replaced as soon as possible.
- Items such as pipettes will be sealed in packets or cans; when removing a pipette ensure that the end that will be used for making transfers does not touch anything except the material to be transferred.
- Pipettes must *never* be placed on the bench; after use, place them TIP DOWN in a plastic tub of disinfectant *immediately*.
- All glassware and similar items which have been exposed to test organisms (often known as *discard* in microbiology laboratories) must be disinfected or sterilized.
- Try to minimize the number of times a vessel is opened (to reduce the number of times potential contaminants may enter).

It is important to remember that the success of experiments depends on the way in which the work is carried out. When working in a laboratory, you have both your own safety to consider and also that of others who may be working in close proximity. Safety considerations are dealt with in more detail in Chapter 2, but the main points to be aware of before starting any procedure in the laboratory are listed as follows.

- The purpose of the experiment must be quite clear.
- What needs to be done and how particular manipulations are to be carried out must be understood.
- Appropriate protective clothing must be worn at all times.
- A 'dummy-run' with nonsterile equipment can be carried out to test your skill and give yourself confidence.
- The equipment should be laid out carefully in a logical sequence for use.

- It is essential to know what to do and to deal with any spillage *immediately*. Ensure that the organism (or test material) is maintained in aseptic conditions at all times.
- *All microbiological samples should be treated as potentially hazardous at all times.*
- Cultures must always be *fully labeled* (organism, medium, date, name).
- *All details* must be recorded (a note book should be near at hand but *outside* the immediate working area, to avoid contact with any spillage).

Most laboratories possess hoods for inoculation and aseptic manipulations. There are several designs for these cabinets, with different safety features that require different operating procedures. Make sure you know the capabilities of the one you use; consult Section 2.3 about the matter.

4.1.1 Contamination

It is extremely important to be familiar with the way in which an organism of interest usually grows. When it is first supplied, grow it up, look carefully at the characteristics in the culture medium by eye and the morphological features under the microscope. Check the features with published descriptions and make full, precise records. Always inspect cultures thoroughly when using them so that if contamination occurs it is likely to be spotted straight away. If you have obtained a culture from a collection or from another investigator, it is important to set up additional (stock) cultures soon after receipt. These can be stored and used later if contamination occurs.

If an extraneous organism does enter the system, it is obviously important to identify the source of the problem. It is useful to know if contamination is bacterial or fungal, but a more precise identification can take a great deal of time and effort and is not usually necessary. It is more important to ensure that the problem does not arise again. Time is well spent deciding if the problem lay with a particular batch of medium, was associated with a defective (e.g. cracked) flask or occurred because the culture was improperly handled. It is possible that an autoclave is not operating at the stated pressure value. Very persistent contamination may need more thorough investigation, but usually such events are relatively isolated incidents.

Aerial contamination. Aerial contamination can be a problem, particularly if there are many airborne spores in the vicinity of the working space where culture transfers are carried out. It is relatively easily spotted in cultures where solid growth medium is used, since a scatter of contaminating colonies will be apparent, usually around the edge of the culture. In liquid cultures the identification of contamination can be more of a problem (see below). Such contamination is most easily controlled by avoiding draughts when culture transfers are made and by good aseptic technique. Always prepare a few more cultures than are actually needed for the experiment if there is any likelihood of this problem.

Contaminated inoculum. Sometimes it is difficult to separate microbes in culture and what appears to be a single culture may harbor more than one organism. For example, bacterial and/or yeast cells can easily be concealed beneath fungal mycelium. The virtue of inspecting cultures very carefully, both by eye and under the microscope, is therefore clear. Vigilant inspection of cultures will identify a problem before too much time and effort have been wasted using contaminated cultures. If this sort of contamination is suspected, methods for separating organisms can be employed (Chapter 7), usually to good effect.

Liquid cultures. Deciding whether liquid cultures have become contaminated is a common problem. If there is any suspicion of contamination the culture should be inspected using a microscope and Gram staining (Section 8.3.2) carried out if necessary. An oil-immersion objective is needed to give a high-enough level of magnification for such determinations. If prolonged or repeated culturing is required, periodic checks should be carried out routinely.

The presence of yeast cells also causes the growth medium to appear opaque and cloudy. Under the microscope, x40 magnification is sufficient to identify yeast cells (because of their size) without the requirement for any staining procedures. Yeasts are common contaminants of laboratory media.

Stirred liquid cultures of fungi may contain soft pieces of mycelium (*pellets*) which can be quite large and spherical. Mycelial fragments may also be small and dispersed, giving a more homogeneous appearence to the culture. In either instance, if the culture is allowed to stand for a few minutes, the mycelium will sink slightly in the culture liquid. If no yeast or bacterial contamination has occurred then the culture fluid will be absolutely clear. The presence of a second fungal species in the system must be determined either by microscopic inspection and/or by placing small fragments from the culture on to solid medium for further inspection.

4.2 Physical requirements for growth of microbes in culture

Microbes show great versatility in adapting to different environmental conditions. However, no one organism can survive all possible conditions, and the prevailing environment does have a profound influence on the growth, development and reproduction of individual species. In order to assess the tolerance of individual microbes to fluctuating conditions it is important to understand their general physical requirements. This will also provide information on how conducive conditions may be mimicked in the laboratory.

The growth of microbes is markedly affected by physical conditions, particularly temperature, pH, oxygen and water availability. Conditions conducive to one species may be lethal to another. For the growth of microbes in the laboratory it is important to identify those factors with most influence, and to satisfy the needs of the particular organism.

With solid media, such as agar, there will not be a constancy of conditions underneath a microbial colony. Thus, taking two important factors, there can be a distinct difference in both the pH and oxygen concentration in the medium at the margin of the colony compared with the medium under the middle of the colony. Indeed under the middle, there may be little oxygen.

4.2.1 Temperature

Most microbes have a *minimum* temperature below which no growth occurs, an *optimum* temperature where most rapid growth takes place and a *maximum* above which no growth will occur (*cardinal temperatures*). The optimum temperature is usually near to the maximum and may be governed by the temperatures at which enzymes involved in key biochemical reactions become inactive or denatured. The lowest temperature, especially if this is near freezing, may be governed by the functioning of cell membranes. Temperature effects can have quite dramatic influences on the growth and development of microbes. Each species will have its own particular temperature profile.

Micro-organisms have been grouped according to their preferences, although these often overlap. Cold-loving organisms with a low-temperature optimum (15–20°C) are known as *psychrophiles* and die at higher temperatures. This definition needs to be treated with some caution. Many fungi grow best in this temperature range and many will grow and reproduce very well below this range. The most numerous group of microbes are the *mesophiles* with a mid-range temperature optimum (20–40°C). Thermophiles have high temperature optima (40–85°C), with most in the 40–60°C range. It is interesting that no eukaryotes grow above 60°C although some bacteria (particularly archaebacteria) are found growing and reproducing at temperatures in excess of 100°C.

In the laboratory, most temperatures can be achieved using insulated, controlled incubators. However, because the effects of even small temperature changes can be dramatic, it is important to be aware of fluctuations and cold spots which can occur in incubators, particularly for sensitive species. Additionally, if the door of an incubator is repeatedly opened and shut, the temperature may fluctuate quite considerably. In a busy laboratory it is very important to consider and take action to minimize such changes. One way to do this is to keep your cultures in a box within the incubator.

4.2.2 pH

Most species grow best over a relatively narrow range of hydrogen ion concentration, although some species may be acid tolerant or alkalotolerant, surviving in extreme conditions. Organisms growing and reproducing in relatively acid conditions are known as *acidophiles* whereas those thriving in alkaline conditions are known as *alkalophiles*. Unlike temperature, the optimum pH is usually near the middle of the pH range for the organism. The pH range that can be tolerated by microbes is highly variable, but most microbial growth occurs between pH 4.5 and 7.5. In general, fungi grow better and are more tolerant of lower pH than bacteria. Algae are also relatively tolerant of pH. Only limited numbers of species are tolerant of very acid conditions, and as a result acidification is used effectively as a means of food preservation.

Active microbial growth in culture may well result in pH change as acidic or alkaline products are released into the growth medium, or acidic and basic ions are depleted from the growth medium. pH changes may be so great that further growth of the organism is prevented. Extremes of pH can affect the availability of nutrients. When there can be a marked pH change it may be necessary to buffer the culture medium. The particular kind of buffer chosen for the purpose will depend on the effects the microbe has, and must be chosen with care. Some buffers can have inhibitory effects on certain species. Additionally, buffering compounds may be metabolized by the microbe concerned.

In some cases, culture medium pH must be adjusted prior to introduction of the organism. It is important to measure the pH value after sterilization, since autoclaving can alter the value.

4.2.3 Oxygen availability

Many organisms require oxygen for growth and development (*aerobes*) whereas others (*anaerobes*) do not use oxygen and may be harmed or even killed by exposure to it (some bacteria, protozoa and a very small number of fungi). *Facultative anaerobes* grow well in the absence of oxygen but suffer no detrimental effects on exposure to air. Some aerobes grow best at lower levels of oxygen than are found in the atmosphere (*microaerophiles*) and some anaerobes (*aerotolerant*) are able to tolerate oxygen but do not use it.

In ideal cultures microbial cells grow and divide very quickly and aerobic organisms often require large amounts of oxygen as a result of this rapid metabolism. It is important to ensure that aerobic cultures are not tightly sealed from the atmosphere (although they must be protected from microbial contamination) and tops of vessels must be left loose for incubation or gas-permeable cotton wool plugs used. The failure of an aerobic culture to develop beyond initial growth may well be due to lack of oxygen. In a closed culture, oxygen levels will rapidly become depleted and levels of carbon dioxide will increase. Even in Petri dish cultures there can be a lack of oxygen, particularly if dishes are stacked in a pile so that the lids are held tightly in place.

Oxygen is poorly soluble in water and is used up rapidly. Where liquid cultures are to be used for microbial growth the vessels used should offer large surface area to total liquid volume ratios in order to provide as much exposure to the available oxygen supply as possible. Static liquid cultures are usually set up as very shallow layers. The diffusion of oxygen into the medium is improved in cultures that are shaken during incubation, by placing on a platform either rotating (rotary) or moving backwards and forwards (reciprocal). In some cases the shaking speed needs to be quite high and, if more oxygen is needed, turbulence can be increased by means of dimples in the sides of the culture vessel. Such vessels are often used for the growth of bacteria or algae. In closed fermenter systems, oxygen-rich air is forced into the system by bubbling (sparging).

It is surprisingly difficult to achieve anaerobic conditions for growth of oxygen-sensitive organisms. It is necessary to seal any culture vessel tightly and effectively to prevent leakage and also to remove all the oxygen which will be present when the vessel is first closed. For strict anaerobes even a brief exposure to low levels of oxygen may be fatal and therefore such organisms require very careful handling, totally within an oxygen-free environment. The isolation and continued culturing of such strict anaerobes requires very specialist equipment and is beyond the scope of this text.

The use of stab cultures (Section 4.3) can help to determine whether any organisms in a sample are anaerobes or facultative anaerobes, since oxygen levels are limited low within agar medium. Aerotolerant anaerobes can be grown in an anaerobic jar, a vessel with a tight seal. After inoculation, cultures are placed into the jar together with a sachet containing chemical agents to evolve hydrogen and carbon dioxide. The hydrogen evolved reacts with oxygen on the surface of a palladium catalyst, forming water and giving rise to anaerobic conditions. Such sachets are available commercially (GasPak®) and are very successful for most laboratory purposes. However, it is vital that the seal is efficient otherwise a suitable environment will not be achieved.

Some organisms, particularly those that are capable of photo-synthesis, may be limited in culture by the lack of sufficient carbon dioxide. The requirement can be met by adding $NaHCO_3$ (± 20 mg 1^{-1}) to the medium.

4.2.4 Water availability

Water is required by all microbes, in varying amounts. Some are fairly resistant to drying and others are fully aquatic species. Culture media are often liquid based, but sugars and salts are also included, sometimes in large amounts. These may make the water in the system unavailable to the microbes. The amount of available water in a system is best expressed either as *water activity* or as *water potential*. The physical principles underlying these two terms are covered in considerable detail elsewhere [1,2].

Water activity is reduced when a solute (e.g. sugar) is dissolved in water; some of the water is associated with the solute and becomes unavailable. The water activity of pure water is 1.0 and goes down to zero when absolutely no water is available. Maple syrup has a water activity of 0.90, jam 0.80 and dried fruit 0.7. A few fungi can survive water activity as low as 0.65. In order to survive low water activity cells reduce their own water activity by pumping ions in from the environment or, as the water activity falls, by the uptake or synthesis of a compatible solute (noninhibitory to biochemical processes), for example polyols, proline. Organisms requiring salty media of low water activity are known as *halophiles,* and those growing well in sugary media of low water activity are termed *osmophiles.* Many micro-organisms are able to grow well in both types of media.

Freezing also lowers water activity, that is, reduces the amount of water available for use by microbes, and has a very important role in the preservation of foodstuffs and other perishable materials. Reducing the temperature of pure water to −15°C lowers the water activity from 1.0 to 0.86. Under such conditions the growth of microbes is restricted (including that of cold-tolerant species) as a result of the reduction in water activity, although it is likely to resume if higher temperatures are restored. Indeed, controlled freezing can be an important method of preserving microbes (see Chapter 9).

Desiccation is a problem for most bacteria and many other microbes too, resulting in membrane damage. However, some species of fungi, particularly the lichenized forms, are very resistant to desiccation (*xerophiles*) and grow in regions where the water supply is limited and/or intermittent.

4.2.5 Availability of light

Some microbes (*phototrophs*) have an absolute requirement for light, and use it as their source of energy for growth (e.g. algae, cyanobacteria, purple and green sulfur bacteria). Light provides the energy source for photosynthesis. Some microbes have pigments which absorb light energy (e.g. chlorophylls, 400 nm and 600–800 nm; carotenoids, 450–550 nm; phycobiliproteins, 550–650 nm), others have pigments to protect the cells from damage by harmful ultraviolet rays. Particular wavelengths of light are sometimes stimulatory to growth and may be used in the laboratory to enhance the culturing procedure.

Most fungi, however, do not require light to grow and will often thrive in poorly lit situations. However, in some cases light is stimulatory to, and indeed can be required for, differentiation processes (e.g. fruit-body formation) leading to sporulation, although light lasting a few seconds is often all that is required. Different wavelengths may be stimulatory to fruit-body formation in different species, from near ultraviolet (250–320 nm) through to red/far-red (550–670 nm). In some species exposure to light and dark cycles of 24 h duration, often coupled with fluctuations in temperature and humidity, lead to entrained rhythmic (circadian) sporulation and also release of spores.

4.2.6 Culture vessels

The primary requirement of culture vessels is to provide sterile chambers for a suitable nutrient medium which can be inoculated subsequently with the test organism prior to incubation. The vessel must therefore be transparent, so that the growth of the organism can be observed directly, and the sterility of the cultures will be maintained most efficiently if access to the vessel is easy.

Most micro-organisms are initially isolated on to nutrient growth medium contained within Petri dishes. When containing nutrient agar and organism(s) these are referred to as *Petri plates*. These are normally round, clear plastic dishes, with a lid that fits well, overlapping the base, to prevent any extraneous microbes entering the culture. The lid is lifted at a slight angle to allow inoculations to be made with minimum contamination (*Figure 4.1*). Petri dishes are usually purchased ready sterilized and growth medium is added in the laboratory. After use they are autoclaved to kill the culture and are disposed of (see Section 2.3).

Glass test tubes, conical (Erlenmeyer) flasks and a variety of glass tubes and bottles with fitting caps are also used. Glass vessels can be cleaned easily and sterilized (see also Section 4.1 for sterilization procedures), both before and after use. Where fitting lids are not part of the vessel, the tops are sealed with cotton wool bungs, the whole top and bung being covered with foil before autoclaving. The foil can then be lifted off prior to inoculation and the vessel opened after flaming, reducing the chance of contamination.

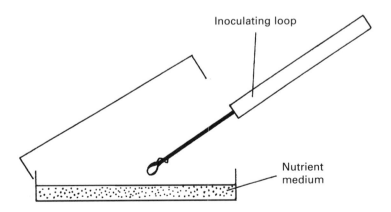

Inoculating loop

Nutrient medium

FIGURE 4.1: *Petri dish lids and bases have high sides so that, when in place, the lid makes a good overlap with the base. The lid should be lifted at an angle to allow entry for inoculum but minimize potential contamination.*

Glass bottles of many different designs can be used for incubation of static (though not shaking) cultures. Providing that the culture medium can be easily inoculated and the vessel will remain sterile during incubation, there is little restriction on the size or shape of the vessel to be used. However, it is usual to fill vessels to only about one-third capacity to allow the presence of air above the culture. For agitated cultures, conical flasks are the culture vessels of choice, though of course these can be used for static cultures too. A relatively small amount of liquid medium is used in each vessel, for example 50–75 ml in a 150 ml capacity flask, so that a high surface area to volume ratio is achieved. When agitated the liquid will be well stirred without wetting the bung.

Manufactured bungs, made from compacted tissue or sponge rubber, for conical flasks and test tubes can be purchased. Cotton wool bungs, made in the laboratory, are very efficient seals if suitably constructed. It is important that the structure of the bung itself is solid and is a good fit for the flask (so that it will not be sucked into the flask during autoclaving) but without being too tight. A strip of cotton wool, about twice the depth of the finished bung and of a length determined by the diameter of the flask neck, is taken from a roll. It is then rolled up under maximum finger pressure against a hard surface. The edges of the torn strip are tucked in at top and bottom as rolling is in progress. Once rolling is complete, slight pressure from top and bottom will mould the bung into its final shape (*Figure 4.2*). Providing that the neck of the flask is clean, cotton wool bungs will give good service for many cultures.

Care must be taken to avoid all cracked or chipped glassware. It might shatter during sterilization or subsequent use and may admit contaminants.

4.3 Selection of culture system

Aseptic techniques and manipulations are common to all aspects of microbiology, but the means of transfer of cultures and the actual vessels used often differ for bacteria, fungi and algae. It is important to make initial decisions concerning the purpose of the experiment and, therefore the kind of system required, in order that cultures can be appropriately designed and set up. The particular nutrient mixture to be used for isolation and culturing is obviously of prime importance. This is considered in greater detail in Chapter 3.

Basic methods for the routine handling of single-organism cultures are given here. The following two chapters deal with culture procedures for experimental studies.

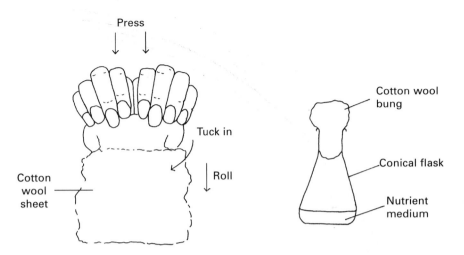

Figure 4.2: *Construction of a cotton wool bung for glassware.*

A single microbial cell provided with conducive conditions will grow and reproduce to form a _colony_. Every cell in an inoculum, which is capable of growth, has the potential to form a colony. All agar plate cultures are incubated UPSIDE DOWN to prevent moisture (condensation) falling on to the culture surface and disrupting any colonies. Every Petri plate in use must be labeled on the *bottom* section of the plate (that section containing the agar and the organism). Condensation must be kept to a minimum and it is common practice to surface dry agar plates before inoculation by leaving them, with the lids off, for a short time (15–30 min) after pouring. This *must* be done in a laminar air-flow cabinet otherwise the medium will become contaminated.

Streak plate method. The streak/spread plate method (*Figure 4.3*) is most usually used for the isolation of *single colonies* of bacteria. In an inoculum, every cell that is capable of growth (*colony-forming units*) will form an individual colony. This technique is a useful aid for separating individual cells, and the resultant colonies can be picked off for subculture.

- An inoculating loop is sterilized by holding in a Bunsen flame, at such an angle that the *whole* wire glows red, and is then allowed to cool in air (*Figure 4.3a*).
- The sample is uncapped, the neck of the vessel flamed and a loopful of inoculum removed (*Figure 4.3b*).
- The vessel neck is flamed again and the cap replaced.
- The inoculum is streaked on to the surface of the agar as shown in *Figure 4.3c. Remember that the inoculating loop must be flamed between each streak made.*

(a)　　　　　　　　　(b)　　　　　　　　　　(c)

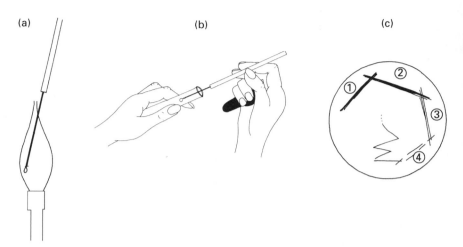

FIGURE 4.3: *Streak plate method. (a) Thoroughly sterilize the inoculating loop by holding in a Bunsen flame. (b) Uncap the sample, flame the top of the tube and take up some inoculum on the inoculating loop, reflame the top of the tube and replace the top. (c) Streak the inoculum, spreading it on to nutrient medium in a Petri dish; then flame and cool the loop before streaking again as indicated.*

- The inoculated agar plate is then incubated, upside down, in appropriate conditions.

After incubation, bacterial growth will be seen on the agar surface. Where the initial streak was made bacterial growth will appear as a smear (confluent growth) and individual colonies will not be visible. Further streaks have the effect of diluting the inoculum and at the position of the last streak (streak 4 in *Figure 4.3c*) individual colonies will be apparent. Using aseptic techniques and an inoculating loop, it is then possible to pick up a small portion of an individual colony and place it on to clean agar. In this way a mixed culture can be separated into single colonies and grown up individually. Streak plating has the effect of diluting a culture.

The method can also be used effectively where fungal growth might be expected. In this case, however, owing to the spreading nature of hyphal growth, the cultures can quickly become overgrown with mycelium and the presence of some bacterial colonies, particularly any that are slow to develop, may be masked rapidly. It may be necessary to use a relatively small inoculum and make sure that it is very well spread. When discrete colonies reach manageable size, it is advisable to make subcultures.

Spread plates. Spread plates can be used to separate colonies (or to estimate numbers of microbes in a sample after dilution of the original; see Section 5.2.2). This method can be used for mixed cultures and for fungi although, again, bacterial colonies may be overgrown by fungi if long incubation times are needed. Using aseptic technique:

- the sample is appropriately diluted with sterile distilled water (the aim is to place fewer than 100 colonies on a plate);
- a known aliquot (0.1 ml) of several different dilutions of the sample is pipetted on to the agar. At least three plates for each dilution are required. The water carried over in the aliquot used will be absorbed into the agar fairly quickly but it is important not to use more than 0.1 ml or there may be problems with surface moisture;
- a sterile glass spreader (bent glass rod) is used to spread the sample over the agar surface (see *Figure 5.3*);
- plates must be incubated under appropriate conditions for the organism in question (these plates must not be turned upside down until the surface moisture has been absorbed).

After incubation, bacterial colonies will be seen and individuals can be picked off for further subculture.

Seeded plates. Seeded plates can be used to test the sensitivity of a micro-organism to a compound (e.g. an antibiotic). A 'lawn' of confluent bacterial growth is required, to cover the whole agar surface. This is achieved as follows:

- an aliquot of inoculum is placed (seeding) into 15 ml molten medium (held at 40°C, above setting point for the agar but not too hot to kill the organism) in a glass vial (Universal bottle);
- the contents of the vial must be mixed well by rolling the bottle between the hands;
- the inoculated medium is then poured into an empty, sterile Petri dish;
- small filter paper disks soaked in test compound are placed directly on to the agar surface;
- the inoculated Petri dishes are then incubated as appropriate;
- in regions where microbial cells are unaffected by test compound confluent growth will result (opaque areas);
- in regions where microbes are inhibited by the test compound no growth will occur (clear areas).

An alternative first step is to place an aliquot of inoculum directly into the Petri dish and then add sterile molten medium to the dish. In this case the dish must be rotated to mix the culture and the medium so that the inoculum is evenly distributed. Where clear zones of inhibition are seen around test disks the relative degree of inhibition of different compounds can be estimated.

Seeded cultures are often used for testing the sensitivity of bacteria and algae to chemicals. However, the same technique is used for a different purpose with fungi. The use of liquid cultures for production of fungal biomass often requires large amounts of inoculum. The medium is seeded with hyphal fragments or spores to produce relatively uniform growth across the medium. So with a seeded culture a lawn of spore-producing hyphae is produced; the spores can be washed off with a small quantity of sterile distilled water. In seeded fungal cultures when hyphae reach the agar surface spores will often be produced. This is a reliable means by which large numbers of spores may be obtained, of relatively uniform age and physiological status.

Fungal subcultures. Cultures of filamentous fungi grow out radially from a point inoculum, giving rise to a circular colony of mycelium. At the outer margin of the colony hyphae will actively spread into fresh nutrient medium and will represent the youngest growth; towards the center of the colony aerial mycelium will be formed and in older regions, towards the center, sporulation may occur. The observed differences are largely attributable to age; both age in time and physiological age (see Section 5.3 for a detailed discussion of mycelial growth).

Inoculation of fresh medium from a mycelial colony can be achieved by using either the transfer of spores or transfer of mycelial fragments.

Asexual spores (particularly conidia) are usually formed in mature regions of the colony and these may be picked up, using an inoculating loop, and placed on to fresh agar medium. Often such spores are very dry and easily dropped from the loop. This leads to the formation of very scattered colonies. Spores are better placed into water prior to inoculation to avoid this problem.

Hyphal fragments can be transferred using an inoculating loop or by using a small piece of agar, with mycelium, cut from the original culture and placing on to fresh medium.

Slope cultures. Petri dishes are ideal for relatively short-term cultures. If slow-growing organisms are to be used, or if cultures are to be stored for any period, then it is more usual to use slope cultures. Agar medium is set, at an angle, into glass vials (or sometimes bigger bottles) to give a large surface area for growth. Inoculum is introduced and the cultures incubated as usual.

These cultures are also used for the production of very large numbers of fungal spores, since suspensions can be easily collected from a well-stoppered bottle. An aliquot of suspending solution (water or nutrient medium) is aseptically pipetted on to the culture surface and the bottle top replaced very tightly. The liquid can then be rolled or gently shaken over the surface of the culture, washing spores off the culture surface and into suspension. It may be necessary to estimate the numbers of spores per ml suspension fluid using a hemocytometer (see Section 5.2.2) and appropriate aliquots can then be used as inoculum.

Stab cultures. Stab cultures are used in order to evaluate the oxygen sensitivity of microbes in a natural environment. Agar medium is set into a test tube to give a column of agar 7–8 cm deep (*Figure 4.4*). A *straight* metal wire is used in place of an inoculating loop.

- Using aseptic technique throughout, some test organism is picked up on the wire.
- The wire is then inserted into the center of the agar in the tube and stabbed to the bottom of the tube.
- The tube is then incubated appropriately.
- Growth at the surface only indicates an aerobic organism.
- Growth only deep in the agar indicates an aerotolerant anaerobic organism.
- Growth throughout the agar indicates a facultative anaerobic organism.

Culturing pathogens. Symptoms typifying a particular disease are often accompanied by the presence of a micro-organism. From direct observation of a wound or infected tissue it may not be possible to determine if any micro-organisms present are the natural flora of the host or if they are causing the disease symptoms. In 1876 Robert Koch, working with animals, established a procedure to determine whether a particular microbe is directly the cause of a disease. Koch listed certain requirements which must be fulfilled before a microbe can be regarded as a pathogen. The procedure is summarized in Koch's postulates, listed as follows.

- The micro-organism must be consistently associated with the same disease symptoms.
- The micro-organism must be isolated into pure culture and its specific characteristics studied.
- Characteristic symptoms of the disease must develop after re-inoculation into a healthy host.

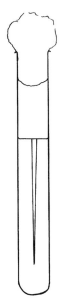

FIGURE 4.4: *Stab culture. Inoculate centrally deep into the agar.*

- The micro-organism must be re-isolated from the test host and identified as that originally isolated.

These rules are still valid today and apply equally to pathogens of plants, animals and other organisms, although we now appreciate that it is sometimes difficult to satisfy these exacting requirements in certain cases. Notable exceptions are those pathogens that cannot be grown in culture (e.g. viruses).

In the laboratory there are stringent regulations concerning work with some potential pathogens (particularly human pathogens) because of health considerations (see Chapter 2). Nevertheless, studies of other pathogens can be possible under normal laboratory conditions. Such study is of increasing importance, since pathogens may be sources for the biological control of pest organisms.

References

1. Griffin, D.M. (1981) *Adv. Microb. Ecol.,* **5,** 91–136.
2. Jennings, D.H. (1990) in *Microbiology of Extreme Environments* (C. Edwards, ed.). Open University Press, Milton Keynes, pp. 117–146.

5 Growth of Microbial Cultures

5.1 Introduction

A bacterium, a yeast or a unicellular alga grow by increasing the number of independent cells. A filamentous bacterium (actinomycete) or fungus grows by an increase in the volume of the colony (mycelium). This chapter is concerned with the different methods that are required for determining, for the two types of micro-organism, the amount of growth that might have occurred. Additionally, one may wish to know how many of the cells or how much of a mycelium is metabolically active or capable of continued growth. Appropriate methods are also considered here.

5.2 Increase in cell mass and number

Growth of a suspension of cells can be determined as an increase in mass or cell number. Most methods tend to ignore the fact that a culture of cells is heterogeneous. Any estimates of growth are averages. Even though individual cells may be identical genetically, they differ one from another in terms of their age, dimensions, antibiotic resistance and wall composition. These differences are brought about by the increasing asynchrony as the culture grows. So any property of a culture will, if measured, be an average value for the culture under the particular given conditions.

5.2.1 Cell mass

Collection of cells by centrifugation. The wet or dry mass of a culture of cells is determined by the difference between a centrifuge tube containing the cells and the tube preweighed prior to harvesting.

- The suspension of cells is centrifuged in a calibrated high-speed centrifuge tube for a time and at a speed such that the cells are sedimented and the supernatant liquid decanted with minimum loss of cells. It will be necessary to carry out a preliminary investigation to determine the most appropriate speed and time.
- Before the supernatant liquid is decanted, the volume of the packed cells should be determined.
- The mass of the wet cells and the centrifuge tube is then determined. If the glassware is properly clean there should be no drops of liquid on the side of the tube but, if there are, these should be wiped off gently before weighing.
- The tube and cells are then left in an oven at a little above 90°C until constant mass is achieved. Before each weighing, the tube must be cooled to ambient temperature in a desiccator.

What problems there are with the above procedure revolve around the presence of culture medium remaining associated with the cells after centrifugation. This has two consequences. First, the volume of sedimented cells measured as above must be corrected for the volume of liquid between the cells (interstice volume) before a value can be obtained for the total cell volume. Secondly, in similar manner, the value for the total mass of sedimented material must be corrected for the mass of medium in the interstice volume. A mass of rigid spheres contains an interstice volume of 27%. With living cells, the interstice value may be similar in magnitude or less, depending upon the shape of the cells and their deformation during centrifugation. Whatever the procedure used to determine interstice volume and therefore the mass of solute contained therein – and such procedures are described in that part of Section 5.3.2 concerned with protoplasmic volume – it will be obvious that a set procedure of centrifugation is required to ensure that the degree of deformation does not differ significantly between each determination. When the interstice value is obtained, the mass of the culture medium in the packed material can be calculated, given that the mass of the medium per unit volume is known. When dry mass alone is being determined, the problem of culture medium being retained in the centrifuged mass of cells can be obviated by washing the suspended cells with distilled water and recentrifuging one or more times before the final centrifugation. Remember that washing with water can lead to loss of material from cells if they burst due to an unfavorable osmotic gradient, and indeed the same can occur even from intact cells. Such possibilities need to be checked before washing is used routinely.

Collection of cells by filtration. Collection of cells by filtration under gravity or vacuum is the customary method for determining their mass free of interstitial liquid. Bacteriological filters (see Section 4.1) should be used. Check that the type that you plan to use does not lose mass after washing and drying.

5.2.2 Cell number

Direct counting. This is performed with a hemocytometer, which was originally designed for counting red blood cells. It consists of a thick glass slide with a central rectangular platform between two deep, transverse grooves (*Figure 5.1*). The surface of this platform is optically plane and at its centre is engraved a 1 mm square grid, subdivided into 400 smaller squares. An optically plane coverslip is applied so that its ends are supported by the surface of the slide bordered by the outer edges of the two grooves. When the slip is correctly positioned, there is a gap of exactly 0.1 mm between the underside of the slip and the surface of the central platform. A drop of the cell suspension placed on the edge of the coverslip will run on to the grid and fill this gap by capillary action. The number of cells lying within the 1 mm square grid can then be counted under the microscope. This will be the number in 0.1 mm^3 (0.1 µl). Since there are 1000 mm^3 in 1 cm^3, multiplying by 10^4 will give the number of cells cm^{-3}. The steps of the actual procedure are listed as follows.

• The slide and coverslip are washed, finally with distilled water, and blotted dry. Then they are wiped with a little alcohol to free the surfaces of grease, and polished with lens tissue. This whole procedure is repeated before and after each sample.

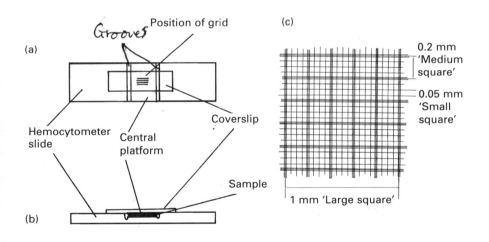

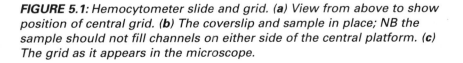

FIGURE 5.1: *Hemocytometer slide and grid. (a) View from above to show position of central grid. (b) The coverslip and sample in place; NB the sample should not fill channels on either side of the central platform. (c) The grid as it appears in the microscope.*

- The slide is placed on the bench and the coverslip lowered into position. The edges of the coverslip are then pressed with the thumb and at the same time a slight sliding movement is imparted. Colored bands (Newton's rings) will appear where the coverslip is in contact with the slide, which means that the gap between the slip and the platform will be exactly 0.1 mm. The middle of the coverslip, where it is not supported, must not be pressed.
- The cell suspension is introduced, by means of a Pasteur pipette, to the edge of the coverslip. It is important that sufficient liquid is allowed to flow under the coverslip to fill the gap between it and the platform. Liquid must not be allowed to enter the grooves on either side of the platform. Before taking each sample the culture must be mixed thoroughly, in order to have a uniform suspension of cells.
- The slide is placed on the microscope stage and the medium-power objective lowered as far as possible over the coverslip. If you are worried about lowering too far, look at the distance between the objective and the coverslip from the side of the microscope. Raise the objective lens slowly, looking down the eyepiece until the grid is in focus. The grid is delimited as shown in *Figure 5.1c*.
- There should be 5–15 cells in each smaller square. If this is not so, and this is usually the case, the suspension will have to be diluted until the requisite number of cells occurs in each square.
- The cells in the sample are counted. It is customary to count 50–100 smaller squares. The squares that are counted should lie on the diagonals of the grid. Some cells lie on the boundary between squares. Include in the count only those cells that lie on the top and right-hand edges of any square.
- The whole procedure should be repeated two or three times. Marked variation between counts is almost certainly due to failure to produce a homogeneous suspension of cells.

The mean number of cells counted per smaller square is calculated as follows. The volume above each square is $1/400 \times 0.1$ mm^3. If N is the mean number of cells per smaller square, then $N \times 4 \times 10^6$ is the number of cells cm^{-3}. To obtain the number of cells in the original suspension, any dilutions that have been made must be corrected for.

Optical methods. The amount of light scattered by a cell suspension can be proportional to its concentration, as expressed by numbers of cells per unit volume. Before discussing the technical details of the method of estimating cell numbers by light scattering, it is necessary to give a little of the physical background to the method. When a beam of monochromatic light enters a culture, the quantity that passes through is reduced by absorption and scattering. If the cells are nonpigmented, absorption is negligible. Therefore, in this case, scattering by reflection and by diffraction within the cells reduces the transmission of light through the culture. The total amount scattered increases directly with the ratio of particle size : wavelength of the incident light. This means that scattering will be greater: (i) at a particular wavelength, the larger the suspended particles; and (ii) the shorter the wavelength of the incident light.

Given the appropriate instrument, one can measure the amount of transmitted light or the amount of scattering. Consequent upon the light being scattered, the suspension of cells appears turbid to the naked eye. Thus, instruments measuring light transmission through, or scattering by, a culture are called *turbidimeters*. In principle, the amount of transmitted light can be measured by most colorimeters and absorptiometers acting as spectrophotometers. However, it needs to be realized that most of the scattered light travels in a direction close to the original beam. This means that the more effective instruments are those in which the beam reaching the measuring photocell has been collimated to a very narrow width.

The larger the number of suspended particles, the less light will penetrate the suspension. At low turbidities, there is a simple geometric relationship between cell number and the unscattered light. This is because at low cell concentrations the Lambert–Beer law holds, namely:

$$I = I_0 \cdot 10^{-\varepsilon lc},$$

where I_0 is the amount of incident light; I, the amount of transmitted (undeviated) light; ε, the extinction coefficient; l, the path length across the suspension; and c, the cell concentration. It follows from the above equation, that

$$\log (I/I_0) = -\varepsilon lc,$$

$\log (I/I_0)$ being termed the absorbance, optical density or extinction, and when plotted against c gives a straight line of slope εl.

As the turbidity increases, there is increasing deviation from the Lambert–Beer law due to secondary scattering, that is, the light scattered by one cell is rescattered by another cell. This rescattered light may be directed to the photocell.

There are instruments that measure the light scattered at 90° to the beam which is incident to the cell suspension. Such instruments are called *nephelometers*. A spectrofluorimeter may be used as a nephelometer when the excitation monochromator and emission monochromator are set at the same wavelength.

In principle, the absorptiometer/spectrophotometer is the more effective instrument at relatively high turbidities, while the nephelometer is the more sensitive at low turbidities. This is because in the latter case one is determining the amount of light above zero, secondary scattering reducing effectiveness at higher turbidities. The effectiveness of the absorptiometer/spectrophotometer is increased as the amount of light transmitted through the culture decreases compared with that going through the blank, namely the cell-free solution.

Some precautions

(a) Absorptiometers/spectrophotometers.

- Where possible, a single phototube double-beam instrument should be used. The better its stability and accuracy of the wavelength settings, the better the instrument will be for making measurements over a wide range of cell concentrations.
- A simple single-beam instrument can be used if the blanks are alternated with samples and dark-current measurements in a standard manner. Such a procedure minimizes changes in the phototube response.
- When dealing with rod-shaped cells, their alignment changes with time in suspension. So, whatever the instrument, a standard procedure for mixing and taking readings should be established.
- Whatever the instrument, it will have to be calibrated against cell suspensions of known concentration (though you should realize that you can use relative values to calculate doubling times and relative growth rates). Essentially this requires the production of a concentrated cell suspension grown appropriately. A series of dilutions should be made using the growth medium as diluent, but in such a way that cell numbers do not increase during the operation, that is, cell division is stopped or reduced to a very low level. Either the original culture and subsequent dilutions are maintained at 0°C (but there can be problems here since the suspension in the cuvette needs to be brought up to ambient temperature to avoid condensation on the faces of the cuvette), or a protein synthesis inhibitor is used to stop growth. If you use the original growth medium as a diluent, you will need to check that there have been minimum changes in the light-absorbing properties of the medium as the result of organism growth. Marked changes will mean that the light-transmitting properties of the diluted cell suspension will not be altered solely by the reduction in the concentration of cells. Sufficient dilutions need to establish as precisely as possible the shape of the curve following a plot of $\log (I/I_0)$ versus c, since there will be divergence from linearity for the reasons given above.

- The choice of wavelength to make the measurements will be governed by the maximum sensitivity to the light transmitted through the suspension of cells and the amount of light absorbed by the culture medium. For the former, 450 nm is the more appropriate, while at 650 nm the absorption due to the medium (if it is brown-colored) will be minimal.
- An absorption versus cell concentration curve needs to be determined for all growth conditions used, since cell size and shape can vary with the growth conditions.

(b) Nephelometers. These instruments, while being particularly sensitive for small cell numbers, are equally sensitive to the presence of other particulate material. Such particulates may be present in the medium as the result of autoclaving, change in the medium pH and the production of water-insoluble material as the result of microbial activity.

The instrument must be standardized continuously throughout any set of readings.

- The instrument is set at zero with the culture medium blank.
- The instrument is set up to give a fixed response against a standard material. Since one cannot use a cell culture because of the change with time due to growth, an inert standard must be used. Opal glass is the customary standard. Readings are taken against such a standard, which is used also to calibrate the instrument against known concentrations of cells.

Measurement of culture turbidity without sampling. It is perfectly feasible to determine the cell count in a growing culture over time by regularly extracting samples and measuring their turbidity. However, such sampling decreases the volume of the culture and introduces the possibility of contamination. The use of a flask illustrated in *Figure 5.2* allows the reading of turbidity in the culture at regular intervals without the problems just mentioned. Such flasks can be made readily and can be customized for those spectrophotometers that take a tube.

- When a reading is being taken, the flask must be turned carefully from the vertical (so that the liquid in the flask does not come in contact with the bung) until the tube can be inserted in the measuring instrument.
- All flasks must have free-draining tubes, so no culture liquid remains in the tube after the reading has been made.

The Coulter counter. This instrument enumerates cells electronically. It can be used with great success for counting blood cells but its use is fraught with difficulties when used for enumeration

FIGURE 5.2: *Erlenmeyer flask adapted for direct monitoring of the turbidity of a growing culture of cells, using a spectrophotometer.*

of microbial cells. The principle of the instrument is as follows. Culture medium containing cells is forced through a very small orifice. The size of the orifice is such that its electrical resistance is very high compared with the bulk resistance of the medium. When a particle is moved through the orifice, the change in electrical resistance can be sensed and converted into a countable pulse. Pulses can be counted and sorted electronically. Thus a profile of particle number and sizes can be constructed. If you plan to use a Coulter counter with micro-organisms, you need to be aware of the following four major problems:

- some bacteria can be of a size ($<0.4 \ \mu m^3$) which is comparable with turbulence in the orifice;
- there are problems of interpreting the data generated by the counter because of the failure of cells to separate;
- passage of more than one particle through the orifice within a certain short time-span can often result in nondiscrimination between the particles;
- very often the orifice can become blocked.

Altogether, the use of the Coulter counter is not recommended except where the prognosis for obtaining interpretable results is good and sound advice is available on how best to use the instrument.

Viable counts. Determination of cell numbers by the procedures outlined above gives no measure of viability of the cells in the culture. There will always be a debate as to what is meant by a living cell, but for our purposes we can define such a cell as being one that is able to produce daughter cells. It is appropriate here to equate viable with living. Thus a viable cell will produce, given time, a visible colony on agar or a turbid solution in liquid culture. There are two major procedures for obtaining a numerical measure of viability – colony counts and dilution counts. Either procedure leads strictly to a count of viable particles, since cells may be in pairs or chains, and neither procedure can discriminate between the two.

Colony counts. These are made on agar plates and there are a number of variations in the procedure. Two are described here. In both cases, it will probably be necessary to make what is known as a serial dilution of the original culture in order that an appropriate number of colonies are produced on a plate. The exact nature of a serial dilution is considered later in this section.

(a) Spread plates. A relatively small (ca. 0.1 ml) but known volume of the sample is pipetted on to nutrient agar and spread uniformly over the surface. The agar should be of uniform depth and bubble-free. If the agar is too deep, there is less contrast between the colonies which develop and their background. If the agar is not deep enough, the nutrients become limiting and also the agar can dry out. The agar should be dry enough to absorb quickly the water in the culture fluid and any small droplets of condensation on the lid of the dish, which can be shaken off before spreading. The appropriate number of colonies produced on the plate should be around 100–200; for this number to be achieved, the culture may have to be diluted, as indicated above.

- A known volume of culture is pipetted on to the agar and spread with a spreader made out of a Pasteur pipette (*Figure 5.3*).
- The spreader is sterilized before use by dipping the bulk of it in ethanol and flaming it; you will find that only trial and error will produce an even spread over the surface of the agar.

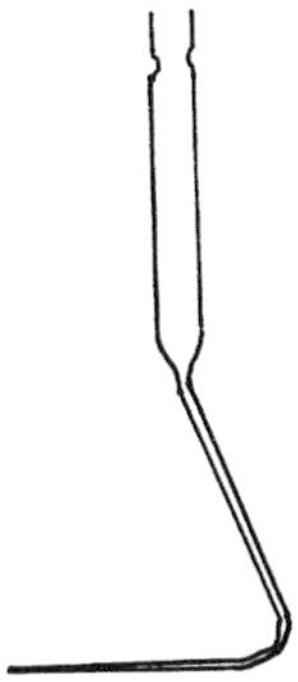

FIGURE 5.3: A Pasteur pipette modified for use as a spreader.

- The plates are inspected before the colonies are fully developed. If the density of colonies is high, then it is possible to make counts with the aid of a binocular low-power microscope. Counting the colonies by unaided eye is facilitated by supplemental illumination, the type and effectiveness of which can only be decided upon by trial and error.
- The colonies should be counted by moving across the plate in a systematic manner. Counting is helped by either using a hand-triggered counter, or by marking the location of the colonies on the bottom of the dish.
- With television-based scanning equipment for counting (and this is to be recommended when counting large numbers of plates), there is a need to check that false-positive counts are not made due to sediment in the medium or dirt on the outer surface of the dish.

(b) Layered plates. This is a somewhat more laborious procedure than that just described, but the colonies produced are more compact and, while a little more difficult to count, because of their size, it is frequently easier to discriminate colonies one from another.

- Prepare nutrient agar plates as just described for spread plates.
- Fill tubes with 2.5–3.0 ml sterile 0.7% (w/v) agar containing nutrients, maintained at 45°C to maintain the agar in the molten state.
- Pipette a small known volume on to the lip of a filled tube and pour the agar over the drop on to the agar in the dish, which is then rapidly swirled to allow the molten agar to cover that already present in the dish.
- When the agar has solidified, pour a further 2–3 ml of the same agar (but without nutrients) on to the agar in the dish.
- Incubate and count as above.

Dilution counts (most probable number method). In this method the culture is serially diluted (see below) and then aliquots of each dilution are pipetted into tubes containing equal volumes of sterile nutrient medium. After a period of incubation, all the tubes that are inoculated with the more concentrated suspensions will be turbid. Those that were inoculated with the more dilute suspensions will be turbid or clear, depending on whether or not the tube received cells. The distribution of cells follows a Poisson distribution. Thus the mean number can be calculated for the particular distribution from the formula:

$$P_0 = e^{-m},$$

where m is the mean number and P_0 is the ratio of the number of tubes with no growth to the total number of tubes.

The dilution method is not very effective in terms of statistics because each tube corresponds to only a small fraction of the area of the agar supporting the colonies. Accuracy must come from the use of many samples. Nevertheless, the dilution method is capable of automation such that many samples can be run. The method does have the virtue over the plate method in that it can deal with cultures with highly variable growth characteristics. Such variability leads to growth on the plates of large colonies which obscure those that are smaller due to their slower growth rate. Also, the dilution method can be used if the organisms produce compounds that form precipitates which can be confused with the colonies.

(a) Serial dilution. Both procedures, plate counts and the dilution method, almost always require that the original culture be diluted to obtain discrete colonies or zero turbidity in some tubes. To do this one carries out serial dilutions, to which reference has already been made. This involves diluting the culture by a series of constant increments. If we appreciate that a culture can contain 10^9 cells ml^{-1}, the series of dilutions can be quite large. How many depends on the dilution factor used. A factor of 10 is customary but it may be necessary to enhance accuracy by having the dilutions more closely spaced. When making dilutions, do not allow the samples to stand for long; keep them agitated to prevent cells from becoming adsorbed on the glass.

(b) Statistics. A proper statistical analysis is required to assess the accuracy of the results that are obtained. Almost all statistical texts are written without the needs of microbiologists particularly in mind. So you are advised to look for a text which is appropriately oriented. We think that the best introduction is that by Meynell and Meynell [1]; Koch [2] also gives good guidance.

5.3 Increase in mycelial volume

Fungi and actinomycetes grow in a filamentous manner. The growth of actinomycetes is much slower (because of the size of the filaments) than that of fungi. For that reason, growth of the former organisms has not been so well characterized. In consequence, this section will focus on the quantification of fungal growth. Nevertheless, the investigator should be left in no doubt that there are remarkable similarities in growth form and growth kinetics of filamentous fungi and actinomycetes, in spite of their very different evolutionary positions [3].

When a fungal spore germinates, the extending hypha grows exponentially for a short time, after which growth of the hypha decelerates. This deceleration is associated with the formation of a branch. If one measures the total hyphal length of the growing colony, it is found to increase in a logarithmic manner with time. The number of hyphal apices and the total dry weight of the colony also increase logarithmically. These observations indicate that there is a unit of hyphal growth which consists of a hyphal apex and a constant mean length of hypha. Thus:

$$\frac{\text{Total hyphal length}}{\text{Number of hyphal apices}} = \text{a constant.}$$

This constant is the *hyphal growth unit* and can be considered as equivalent to the single microbial cell. The doubling time, which one can obtain from a plot of log dry mass of fungal mycelium as a function of time, is the time taken to *duplicate* the hyphal growth unit.

At the very early stages of colony development it is possible to determine the number of units unambiguously because one can measure the total hyphal length and count the number of hyphal apices [3]. But, as the colony develops, in a very little time it becomes impossible to make such a determination because of the density of the hyphae. Also, it is not difficult to see that, because a majority of the hyphal apices are at the margin, the proportion of the colony devoted to hyphal growth units will decrease as the colony grows. The situation is described semiquantitatively in *Figure 5.4*. Thus the deceleration phase which becomes apparent when the log of the radius of a fungal colony is plotted against time represents the stage when a proportion of the colony is no longer involved in active growth.

That proportion of the colony involved in colony radial extension, namely K_r, is called the *peripheral growth zone, w*. In which case

$$K_r = \mu w,$$

where μ is the specific growth rate. It is clear therefore that with: (i) a relatively large colony on agar (in which the peripheral growth zone is a small proportion of the total colony), and if (ii) the colony is assumed to be planar (the production of aerial hyphae is small compared with the number of those spreading outwards over the agar), then the following holds:

$$r = w\mu t + r_0,$$

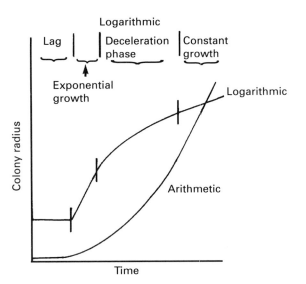

FIGURE 5.4: *A semiquantitative representation of the growth of a filamentous fungal colony on the surface of agar.*

where r is the radius after time t and r_0 is the radius at time zero. If r becomes much greater than r_0, r will be a linear function of t. So for comparative purposes, the most effective measure of the growth rate of a colony is when it is a linear function of time. Of course, it must be remembered that a difference in growth rate brought about by different conditions, for example nutrient concentration or temperature, may be due to a change either in the specific growth rate or in the width of the peripheral growth zone.

All the above indicates that, as far as fungi are concerned, there is a need for a different concept of viability from that considered with respect to a suspension of microbial cells. Thus, if a young colony were to be fragmented (see Section 5.3.2, concerned with growth of mycelium in liquid culture), the great majority of fragments produced would be able to produce a new colony if put on to nutrient agar. Many of these fragments would come from parts of the colony away from hyphal apices and also away from the protoplasm contributing to the hyphal growth unit.

The essential point is that as a colony grows it becomes metabolically differentiated. In some fungi the colony shows morphological differentiation, the production of either new types of hyphae, for example larger diameter, (if septate) larger compartments, thicker walls or spores or other reproductive structures. A striking example of metabolic differentiation can be seen in colonies growing in agitated liquid culture. Under such conditions, growth takes place in all three dimensions and the mycelium is essentially spherical (forming what is known as a *pellet*). When the dry mass is measured, the initial growth of the pellet is exponential but it is not long before there is a deceleration in the increase in dry mass. This deceleration is due to the bulk of the colony no longer contributing to the increase in volume of the pellet. Also, as the pellet grows, there will be an increasing difficulty for oxygen to reach the center of the colony. This will be particularly so if the hyphae are packed very close together, such that the medium in the space cannot become agitated, and thus the gas can only move to the center of the colony by diffusion. When the hyphae are more loosely packed, the liquid in the spaces between the hyphae can be made turbulent, allowing the more rapid convective (bulk) flow of oxygen into the center of the pellet.

The center of large pellets (0.5 cm diameter) can become devoid of hyphae as a result of the anaerobic conditions which develop as a consequence of the inability of oxygen to reach the center from the bulk external medium. If mycelial pellets are being used for physiological or biochemical experiments, they should have diameters of less than 0.2 mm. The use of such pellets ensures the aerobic functioning of all the hyphae. There is a further benefit of using pellets of this size. Since not all parts of the pellet are growing exponentially, the proportion not so doing will be much less than in a larger pellet. This means that the results obtained will more closely reflect what might be occurring in exponentially growing hyphae.

Finally, it is a common assumption that filamentous fungi grow in a uniform manner when in culture. Such uniform growth is expressed as a circular colony on agar. However, this is not always the case. *Figure 5.5* shows three examples of colonies growing in a nonuniform manner. The occurrence of sectors resulting from a particular hypha (and all subsequent hyphae originating from it) at the mycelial front, exhibiting a new growth rate or a different morphology (*Figure 5.5a*) indicating genetic changes in a colony is infrequent, and plates with colonies showing sectors should be discarded. Having said that, do remember that sectoring cannot be detected in liquid cultures. Point growth (very rapid extension of a few hyphae from the colony on to the agar in an apparently organized manner; *Figure 5.5b*) and formation of rhizomorphs (what can be considered as a special case of point growth in that a discrete linear organ is produced which can extend at a rate faster than the undifferentiated mycelium; *Figure 5.5c*) tend to be observed much more frequently in basidiomycete colonies. Both are processes of differentiation and are triggered by environmental change, unlike the process of sectoring which is the result of genetic change.

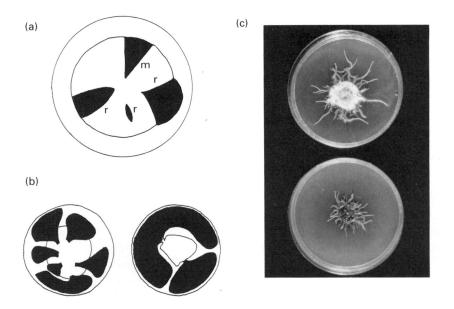

FIGURE 5.5:   *Nonuniform growth in a filamentous fungal colony growing on agar. (**a**) Sectoring brought about by a change in the growth rate (r) or morphology (m) of a particular hypha and all those hyphae that arise from it due to genetic change. (**b**) So-called 'point-growth', in which faster-growing mycelium seems to grow out in a relatively organized manner from 'points' in the original colony. (**c**) Rhizomorphs, linear organs with an organized structure, produced from the mycelium of* Armillaria mellea.

5.3.1 Growth rate of colonies on agar

Linear growth rates on agar are readily determined, particularly with a colony growing in a regular manner.

- Measurements are made at known intervals along two predetermined axes (drawn on the bottom of each Petri dish before the experiment commences) at right angles to each other.
- If the colony grows in a nonuniform manner, the area of the colony has to be measured. This can be done by marking out the outline of the colony on the bottom of the dish at each time interval and at the end of the experiment transferring the outline to tracing paper. The enclosed area can be cut out and weighed, the weight converted to area through having weights for standard areas of paper.
- The agar in each dish must be at a constant, uniform depth.
- Care must be taken in the inoculation of each plate. If a spore suspension is used, it should be of a small volume such that the

liquid present is readily absorbed by the agar. If the inoculation is from a plate culture, small portions of the growing margin should be taken with a sterile inoculating needle or sharp knife.

- Petri dishes should be randomized such that treatments are spread around the incubator (to minimize any temperature variation).
- When the appropriate time comes, the dish must not be out of the incubator for too long a period. This is particularly important if growth determinations are being made at frequent intervals. To speed up the process, the position of the margin can be marked on each axis and the measurements made at leisure at the end of the experiment.

Linear growth rates on agar can often provide useful information about the ability of a fungus to utilize particular nutrients and to withstand osmotic stress, and the temperature tolerance of the organism. The amount of agar (ca. 15 cm^3) present means that it is a culture of limited volume and, because of this, nutrients can be depleted rather rapidly and the pH change quite markedly. Indeed, seeming lack of growth with ammonia as the nitrogen source in certain instances has been shown to be due to the rapid production of an acid pH and not due to an inherent inability of the fungus to use the compound.

It is possible to determine the dry mass of mycelium growing on agar if inoculation is on to a film of cellulose overlaying the agar. Inoculation in these instances is best done with a cylinder of agar and mycelium cut from the margin of the inoculum culture in a Petri dish. At the end of the growth period, the cellulose film together with the mycelium is taken from the dish and the inoculum plug removed. The mycelium is stripped from the film, placed on a piece of weighed aluminum foil and dried at 90°C to constant weight.

Peripheral growth zone. The width of this zone is determined by the following very simple procedure.

- Using a sharp, sterile knife sever the edge of a colony as shown in *Figure 5.6*. The cut should not be made to the base of the Petri dish. Certain hyphae will be cut close to the colony margin, others at an increasing distance from the margin.
- Make a note of the exact position of the colony margin.
- After a suitable period examine the colony. Move along the margin from the cut and examine where the margin has continued to extend. At the point where the distance from the old margin to the new margin becomes constant, the peripheral growth zone is given by the distance from the *old margin to the cut*.

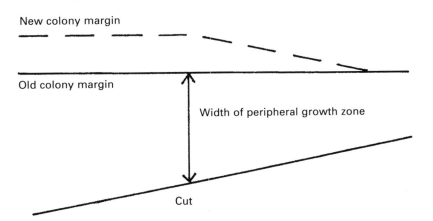

FIGURE 5.6: *A diagram (not to scale) illustrating the determination of the width of the peripheral growth zone for a filamentous fungus. The width is given by the distance between the cut and that point on the original colony margin at which its extension has become unaffected by the cut. Note that an actual colony would have a much more uneven margin.*

5.3.2 Growth of mycelium in liquid culture

Mycelium is grown in liquid culture to obtain biomass grown under conditions that are better defined than those of an agar plate, particularly with respect to the monovalent cation concentration (which can vary with batches of agar), pH (which can be buffered more effectively), homogeneous nutrient concentrations (through agitation of the culture) and aeration (again due to agitation of the culture). Also larger volumes of media can be used.

There are two procedures for producing an inoculum, *spore suspension* and *mycelial disruption.* With a fungus that produces large quantities of spores, a spore suspension can be obtained readily by agitating either a seeded plate culture (see Section 4.3) or a slope culture in a screw-topped tube (see Section 4.3) with a set volume of sterile, distilled water for a defined period. It may be necessary to add a very small quantity of a surface-active agent, for example Tween, to make the spores wettable. For experiments that need to be replicated, it will be necessary to estimate the concentration of spores which have been released and to dilute the suspension to produce the concentration chosen for the inoculation of the experimental media. The concentration of spores is estimated using an hemocytometer (see Section 5.2.2).

The method of mycelial disuption should certainly be used for mycelia that do not produce spores readily, if at all. In fact, disrupted mycelium is a satisfactory inoculum in its own right. The procedure follows from the fact that a high proportion of fragments, produced by disrupting mycelium by relatively short homogenization, are capable of growth such that an extending hyphal apex is produced. There are several steps in the production of such an inoculum, listed as follows.

- Flasks containing liquid medium are inoculated with fragments of mycelium obtained from the growing margin of mycelium growing on agar.
- After a suitable period of agitation, several pellets of variable size are produced within the flasks. The contents of a number of such flasks are poured into a sterile, large-volume (ca. 1 liter) homogenizer, for example, a Waring blender. The pellets are allowed to settle and as much as possible of the medium decanted (carry out the operation as quickly as you can so that the possibility of contamination is reduced to a minimum; it does not matter if there is some medium remaining in the homogenizer). Depending on the number of flasks used and the volume of the homogenizer, it may be necessary to decant more than once.
- When the contents of an appropriate number of flasks (about 5–6 can be suitable) have been dealt with in this manner, 50 or 100 ml of medium is added to the homogenizer using a sterilized measuring cylinder.
- The mycelium is then homogenized at full power in short bursts of 1–2 min, such that the temperature does not rise above ambient.
- Then, using a 5 or 10 ml sterile pipette with the tip removed (unless the tip is removed, the pipette will become clogged) and with the opposite end containing a cotton wool filter, transfer aliquots of the mycelial suspension to flasks containing fresh media.
- These new culture flasks are agitated for a known time. The pellets that grow from the fragmented mycelium are then homogenized as above to produce inoculum for experimental cultures. At this stage the pellets can be washed with either medium, distilled water or standard solutions (particularly ones of known osmotic pressure, if there is concern about too drastic a change to the external osmotic pressure due to the washing procedure).

The mycelial fragments in the inoculum can be considered equivalent to spores – and certainly the equivalence is exact for a spore which has produced a germ tube. Further, if all procedures and growth periods are standardized, the method of inoculation will lead to very similar amounts of inoculum being used each time a new experiment is carried out. There is no need to check on each occasion whether the mass of inoculum is the same; an appropriate constancy can be assumed. If there is a continual throughput of experiments, a stock of liquid cultures can replace agar cultures as the starting cultures.

Both procedures for inoculating cultures have the problem of carry-over, namely the transference to the new culture of substances from the medium used to grow the inoculum. In the case of spores they can be centrifuged and washed with sterile distilled water, though this can increase the risk of contamination. In the case of mycelial homogenization, one should practice decanting to maximize the amount of old culture medium poured off. Always leave the medium until the pellets are properly settled and pour off the medium in a steady but continuous stream. Then the decanting should result in the volume of medium remaining being rather small. Note that it is easy to insert a washing stage following decanting of the culture medium (but see below).

With respect to growing fungi in liquid culture, one should note the following points.

- There should be around 100 ml culture medium per 250 ml flask (or that ratio of volume of medium to volume of flask).
- Constant agitation should be applied at a rate that moves the liquid without breaking the surface. Agitation is required not only to keep the medium mixed and oxygenated but also to prevent mycelium from growing on the walls of the flask. This is more likely to occur with spore inocula. Well-cleaned flasks (water runs cleanly down the walls without leaving drops) are an aid to preventing growth on the walls.
- When agitating a large number of flasks on a shaker with removal of batches of flasks at relevant time intervals, remember that the agitation rate can change due to a change in the loading of the machine. So check the agitation rate of the new loading after the shaker has been set again in motion.
- Unlike cultures of microbial cells, it is rarely possible with fungal mycelial cultures to be able to remove replicate samples from the culture flask over time. With mycelial cultures, experiments must be designed on the basis that at least one sample must be sacrificed for each time point (destructive harvest).

Determination of the mass and other characteristics of mycelium grown in liquid culture.

Dry mass. This is readily measured using bacteriological filters (see Section 4.1) or small-volume (ca. 20 ml) tarred glass vessels with sintered bases of high porosity. The contents of a culture flask can be rapidly poured through the vessel (the medium collected if necessary) and also washed with distilled water or an appropriate solution. The bacteriological filter on which the mycelium has been collected or the vessel with the sintered base plus mycelium should be dried to constant weight at 90°C.

Wet mass. The same procedure should be used for collecting the mycelium. However, when all the mycelium has been collected, a small amount of suction should be applied such that the liquid external to the hyphae and, in the case of the sintered glass vessel, from the sinter itself is removed. Loss of such liquid from the mycelium and the sinter is readily seen and the application of suction immediately discontinued and the filter or vessel and mycelium weighed.

A similar procedure can be used if the pellets are readily manipulated and are capable of being transferred from one container to another. The pellets collected by filtration are blotted very lightly and transferred to glass sinters in 50 ml centrifuge tubes which are then subjected to 100 g for 20 min at 4°C. The mycelium is transferred to preweighed foil and weighed.

Protoplasmic volume. Assuming the density of the mycelium is unity, the difference between the fresh weight and the dry weight gives a measure of the amount of water in the mycelium. This value could be used to calculate the concentrations of metabolites and osmotically active solutes in the protoplasm. However, no allowance would be made for the amount of water in the wall. To obtain a measure of the water so located, the following procedure should be used.

- The pellets should be suspended in a small volume (5 ml) of culture medium or a solution which has the same osmotic pressure as the medium, either of which should contain a compound, at known concentration, which does not cross the plasma membrane of the fungus nor is adsorbed by or bound to the wall.
- After the suspension has been agitated for a suitable short period (2–3 min), the suspension is centrifuged at 1500 g for 5 min. The concentration of the compound in the supernate should then be determined.
- The ratio of the amount (g or mole) of the compound in the 5 ml of initial solution to the amount in the supernate is equal to the ratio of the initial volume of solution (5 ml) to the volume of water in the wall plus 5 ml.

The following compounds have been used: blue dextran and [^{14}C]-inulin; but other compounds that do not penetrate the plasma membrane or bind to the wall can be used. If possible, make measurements with several compounds; this is because it is difficult to assess whether or not adsorption/binding is taking place. Similar values with several compounds point to a correct assessment of the amount of water in the wall. A value with one compound higher than the mean obtained with the other compounds suggests that the particular compound is binding to the wall.

Viability. There is no filamentous fungal equivalent to the viable count for bacteria or yeasts. One can only assess by indirect means the extent to which all the compartments in a fungus are viable. For that reason the concept is rarely (if at all) invoked in filamentous fungal experimental studies. Some indication of the extent to which all the protoplasm in a mycelium is fully functional is given by values of oxygen uptake per unit mass of protein (dry mass should probably be avoided since it can vary with the external environment as a consequence of differing amounts of storage material or changes in the amount of wall). Another possible measure of viability is by stereological examination of electron microscopical sections of pellets. Nonviable compartments are readily identified by the obvious breakdown of protoplasmic components. Stereology allows the determination of the proportion of protoplasm that has broken down [4].

Washing fungal mycelium. When washing fungal mycelium, care should be exercised to ensure that there is neither bursting of hyphal apices, causing significant losses of protoplasm, nor outflow of solutes across the plasma membrane from the intact protoplasm. Bursting is stimulated by unfavorable osmotic gradients, acid pH and non-metabolizable sugars, while calcium ions have a protective effect. So it is preferable to wash with solutions which are isotonic with the mycelial water potential (when grown in normal growth media this will be around -1.0 MPa); 5 mM $CaCl_2$ should also be present in the washing solution. The use of solutions at $0°C$ seems to help prevent bursting and also loss of solutes across intact plasma membranes. It always helps to test the effects of washing solutions on mycelia by analysis of the solution after it has been in contact with the mycelium. The focus should be on classes of compounds or ions which are easily detected, for example UV-absorbing or ninhydrin-positive compounds or potassium.

5.4 A word of advice

Whenever you are using a micro-organism for experimental purposes, make a habit of looking at your cultures with the light or electron microscope. In this way you can maintain confidence that the organism that you are using is genetically unchanged. Ideally, it is valuable to use other tests, genetic and biochemical, though this may not always be possible. As well as helping to ensure that you have cultural stability, microscopy can help in determining how an experimental treatment may be affecting your organism. Remember that quantitative features such as cell or compartment size are as important to focus on as qualitative ones, such as appearance of new inclusions within the protoplast. With respect to determining quantitative features, we must stress the importance of stereological analysis in providing the relevant data.

When carrying out experiments in liquid culture, it is important to perform parallel studies using solid cultures. With such cultures, it can be determined whether your experimental treatments are affecting colony morphology, degree of sporulation and, in the case of filamentous fungi, the degree of branching. Some compounds, the so-called 'paramorphogens' can bring about a dramatic increase in the degree of branching. To the uninitiated, it appears that the growth of the fungus is severely inhibited. In fact, biomass production is almost unaffected.

References

1. Meynell, G.G. and Meynell, E. (1970) *Theory and Practice in Experimental Bacteriology.* Cambrige University Press, Cambridge.
2. Koch, A.L. (1981) in *Manual of Methods for General Bacteriology* (P. Gerhardt, ed.). American Society for Microbiology, Washington, DC, pp. 179–207.
3. Prosser, J.I. and Tough, A.J. (1991) *Crit. Rev. Biotech.* **10**, 253–274.
4. Howard, V. (1990) in *Biophysical Electron Microscopy* (P.W. Hawkes and U. Valdre, eds). Academic Press, London, pp. 479–508.

6 Growth in Liquid Culture for Experimental Purposes

6.1 Introduction

In the previous chapter, we considered the various procedures for determining microbial growth rates on solid and in liquid culture. The amount of medium under consideration is unlikely to be more than 250 ml. In this chapter, attention is focused on growth in liquid culture in much greater volumes, 1 liter or more. There are two reasons for using these larger volumes – to harvest a large quantity of material (either organism or product in the medium) for experimental investigation or to grow the organism under steady-state conditions.

6.2 Large-scale batch culture

The great virtue of growing micro-organisms in large volumes is that one can much more readily harvest at a particular phase of growth. This is because the ratio of inoculum : culture medium volume can be more easily adjusted to extend the length of a particular growth phase. This is especially so for the exponential phase, which can be of very short duration in small-volume cultures. Large-volume cultures are often better buffered against temperature variation in a culture room brought about by frequent entry and exit. On the other hand, as the volume of culture medium is increased there are problems of scale-up. The various issues are considered below. Unless you have a satisfactory answer to each of the questions raised by the issues, you should not proceed. You need to keep in mind that, although there can be virtues in using large culture volumes, it is possible that the comparable ease of handling a large number of small cultures may outweigh these virtues. Even when all the questions have been answered satisfactorily, you may need more than one trial to ensure that you are technically competent with those procedures that you have decided to use. Essentially, the problem is to carry out the necessary operations speedily yet avoid contamination.

6.2.1 Preparing the vessel and the medium

Here are two questions to be answered:

- do you have an autoclave which is large enough for the vessel that you plan to use?
- if you need to filter sterilize some of the nutrients, is it feasible to do so?

Remember that a large vessel cannot be agitated like a small one. This means that the culture medium must be aerated by a supply of air. With respect to this supply, it must be sterilized prior to entry into the vessel. Sterilization is best done with filters – either fiber-glass-packed tube filters or bacteriological filters. In the former case, the packing must not be so tight as to impede the air flow seriously. Bacteriological filters are usually purchased with a custom-designed unit to fit into the air supply. A filter on the air outport will be required as well as the import to prevent entry of contaminating microbes by that route. It is recommended that filters with tubing attached and openings closed with loose cotton wool should be wrapped in aluminum foil and sterilized separately from the vessel containing the medium to avoid the filter becoming filled with water, making it ineffective. Equally, before autoclaving, the import and outport of the culture vessel should be closed with loose cotton wool and covered with aluminum foil. It is important to note that: (i) it is possible to buy prepacked sterile filters, and (ii) there are some investigators who do autoclave the vessel along with the filters attached but with one vent loosely packed with cotton wool to allow rapid pressure equilibration. When attaching the filter to a port of the culture vessel, after both have been made sterile, remove the foil and, if necessary, the loose cotton wool, spray the two connecting pieces with ethanol (glass or metal tube may be flamed) then join the two together. The operation should be carried out quickly, so it is important to make sure that the two tubes fit together easily with an air-tight seal. For ease of operations a quick connector, which allows the glass or metal tube to be readily pushed into the appropriate piece of rubber or plastic tube, should be considered (*Figure 6.1*).

(a)

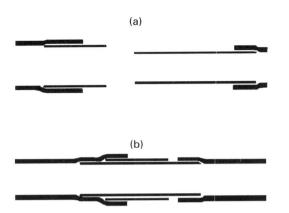

(b)

FIGURE 6.1: *Connector for quick and aseptic connections of tubing among and between reservoirs for continuous culture systems: (**a**) disassembled; (**b**) assembled. Thick tubing, rubber or plastic; thin, stainless steel.*

6.2.2 Aeration

A steady stream of bubbles through the medium is required, of sufficient size and number per unit time to aerate and stir it effectively. The question is:

• do you have the requisite equipment to provide the necessary volume and flow rate of air?

When building your aeration system, make sure that all the joints can withstand the air pressures that you plan to use and, to cope with the possibility that the pressure may vary, ensure that there is a bleed for the system before the air enters the import filter. Aeration causes foaming and you may need to add an antifoaming agent. It is preferable to use a nonmetabolizable agent such as silicone in a 30% (v/v) aqueous emulsion but, whatever the choice, a toxicity test should first be carried out on the micro-organism.

6.2.3 Separation of the organism and medium

Bacterial and yeast cells can be separated reasonably rapidly from a small volume of medium by batch centrifugation. But on scale-up the question is:

• can you deal with large volumes of culture, when the separation process is time consuming, particularly if washing steps are required?

Remember that the longer the time period for the separation the more likely are the changes within both the organism and the medium, making their composition increasingly different from what it was at the time when the separation commenced. With fungal mycelium, the time required will be even longer because of the greater difficulty in getting compaction of the mycelium compared with sedimentation of cells. Indeed, for fungal mycelium it could be better to filter the culture through a wide-diameter sintered glass filter of high porosity. The most satisfactory method for both cells and mycelium, if the culture volume is large, that is, several liters, is continuous centrifugation using equipment made by Alpha-Laval or Sharples. The procedure is rapid and can be carried out at room temperature (or in a cold room if necessary). Batch centrifugation should be carried out in a refrigerated centrifuge, otherwise there will be significant warming over the time period necessary for separation. As soon as possible after separation from the used medium, the organism should be subjected to the required treatment; for example, be treated with an extractant for obtaining the compounds of interest from the organism. Do remember that used medium is almost certainly not organism-free, because it is never possible to separate the solution from the sediment without disturbing a few cells, so either you must take steps for isolating the compound(s) of interest without delay or deep-freeze or filter sterilize the medium.

6.3 Continuous culture

6.3.1 Background

In continuous culture, unlike batch culture, the medium is continuously added, while organism and medium are removed at the same rate. If there is effective mixing, the suspension that is removed is a representative sample of the whole culture volume. Essentially, continuous culture allows continuous exponential growth at a constant rate. When the system functions as described, it is in steady state. What follows below is an introduction to the essential features, both theoretical and practical, of continuous culture. Further information is given in refs 1 and 2.

Let us consider a growing culture of cells. There will be an increase in the number of cells due to cells added and cells produced by growth, while there will be a decrease due to cells being removed and those that die. Thus:

Cells added + Cells produced − Cells dying − Cells removed
= Cells accumulated in the culture.

This equation can be represented mathematically as follows:

$$\frac{FX_0}{V} + \mu X - \frac{FX}{V} - X = \frac{dX}{dt},$$

where F is the medium flow rate into and out of the culture, V is the volume of medium in the culture vessel and μ the specific growth rate (h^{-1}), while X_0 and X are, respectively, the amount of cells coming into the culture and those already present. Of course, if the medium entering the culture is cell-free $X_0 = 0$. If there is a negligible death rate, X can be ignored. Therefore,

$$\frac{dX}{dt} = \mu X - \frac{FX}{V}$$

At a steady state, $dX/dt = 0$; in which case

$$\mu = \frac{F}{V} = D,$$

where D is the dilution rate for the culture.

Putting the above in verbal terms, when the culture is set up the organism grows at maximum rate due to nutrients being in excess. However, cell numbers will reach a value such that one nutrient becomes limiting, that is, it will control the rate of growth, all other nutrients will be in excess, and therefore the culture growth rate decreases to a rate commensurate with the concentration of that nutrient. If new medium is supplied, the growth rate is a function of the concentration of the nutrient and the flow of medium into the culture vessel.

The relationship between dilution rate, D, and the characteristics of the culture is shown in *Figure 6.2*. In that figure, doubling time, steady-state cell concentration, output of cells and the steady-state nutrient concentration are plotted as a function of dilution rate and (as can be seen from the final equation above) specific growth rate. The concentration of cells remains relatively constant over the whole range. The increase in specific growth rate is reflected in the decrease in the doubling time and the increase in the output of cells − there is greater throughput of the organism. The limiting nutrient concentration remains close to zero until the cell concentration starts to fall, due to dilution by incoming medium. Essentially the fall is in that region where $D > \mu_{max}$, namely the dilution is greater than the maximum specific growth rate. That is, the flow into (and out of) the vessel is faster than the organism can grow and divide to maintain its

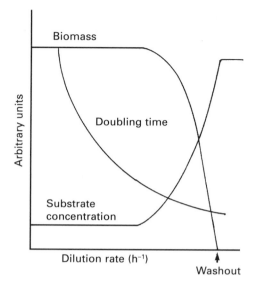

FIGURE 6.2: *Characteristics of a continuous culture of microbial cells as a function of dilution rate.*

concentration. When this happens, the medium coming out of the vessel and the vessel itself become free of cells – this is washout.

There are two major types of continuous culture. The chemostat achieves the steady state by controlling the availability of medium (in theoretical terms the limiting nutrient, as we have just seen), while the turbidostat replaces the cell-containing medium leaving the culture vessel at a rate equal to cell growth. *Table 6.1* compares

TABLE 6.1: *Comparison of chemostat and turbidostat (reproduced from ref. 3 with permission from the American Society for Microbiology)*

Operating parameter	Chemostat	Turbidostat
Operation at or near maximum specific growth rate	Unstable	Stable, very nearly steady state
Operation at low specific growth rates	Stable steady states	Unstable, transient with pulsatile response
Dilution rate equals specific growth rate	Only at steady state	At all times
Cell concentration at constant specific growth rate	Substrate concentration in the feed	Substrate concentration in the feed
Dilution rate	Predetermined	Controlled as a function of cell mass
Substrate concentration for steady-state operation	Requires a single limiting substrate	All substrates may be present in excess

the two types of continuous culture. As can be seen from *Figure 6.2*, the chemostat would be expected to be much less sensitive at or just below the maximum specific growth rate. In practice, the chemostat is used very much more frequently. With the turbidostat it is not very easy to measure turbidity. Cells adhere to the vessel in which their concentration is being measured and air bubbles can also cause problems. For this reason, we shall concentrate on the chemostat.

Figure 6.3 shows diagrammatically the essential components of a chemostat. It consists of a glass fermentation vessel containing the growing microbial culture, which is stirred by an impeller and by sterile air that is forced through the culture. The vessel can be sealed

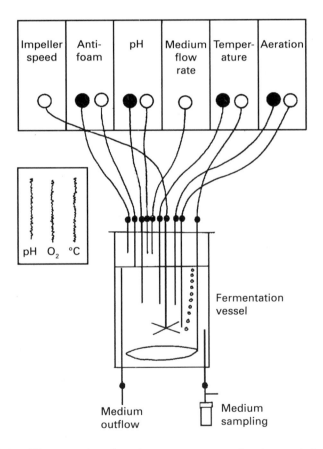

FIGURE 6.3: *Diagram showing the components of a chemostat. On the control modules: •, connection to sensing probe; ○, mechanism for making the necessary correction.*

from microbial contaminants in the surrounding air by a stainless steel lid, with special ports to allow the entry of probes into the vessel without bringing about contamination. Sterile fresh medium passes into the vessel, and during its sojourn there it supports cell growth, while its pH, oxygen content, temperature and capacity to foam is continuously regulated. Medium containing cells resulting from growth of the culture is lost from the fermentation vessel via an outflow tube, at the exit of which the cells and medium can be collected for further analysis. As one will see below, the chemostat is a complicated piece of equipment. Further, a degree of practice is required to operate the equipment satisfactorily. Contamination can often occur in a seemingly inexplicable manner. So if you plan to use a chemostat ensure that:

- there is a good scientific reason for doing so;
- each chemostat run yields as much data as possible.

All this points to good experimental design, and in constructing this it is imperative to plan how you might most effectively analyze the harvested cells and the medium when they are separated. Some important experiments that can only be carried out effectively with the organism of choice in continuous culture (chemostat or turbidostat) are listed as follows.

- The effects of various nutrients when at the particular concentration that limits growth.
- The nature of steady-state growth in the presence of what might be termed 'stressful' agents. Elevated temperature, low and high pH, salinity, toxic metals are obvious candidates.
- The establishment, using a graded series of changes, of how the organism adapts physiologically and biochemically to alterations in the external environment.
- The examination of how two different organisms might compete for a particular nutrient.

With respect to the above, remember that oxygen is a nutrient, and also that comparisons between nutrients containing the same essential element are interesting when the element is in a more reduced state in one nutrient compared to another, for example nitrate versus ammonia. If you are examining the effects of minor elements, keep in mind that contaminants in the other nutrients might change the concentration of the element as specifically added to the medium. In such circumstances, one should analyze the medium for the concentration being used in the experiment rather than that deemed to have been added. Further, because of this problem with contamination, it may not be possible to reduce the concentration to a limiting level. Such a situation can occur when trying to study the effect of a limiting concentration of potassium in the presence of high concentrations (>2 M) of sodium chloride. Even though the contamination of analytical-grade sodium chloride is very small, it can be enough to be greater than the limiting potassium concentration, since sodium itself can replace potassium in some of its cellular functions. Of course, under these conditions, contamination from other sources, notably glassware if not scrupulously clean, can be important.

Limiting concentrations of a particular nutrient are determined by decreasing the concentration until washout occurs. Under such circumstances, the concentration of the nutrient is insufficient to maintain the maximum specific growth rate. Knowing the concentration causing washout, you will need to choose a concentration somewhat higher for your experiments – of sufficient magnitude to allow for the fact that when growing cells at, or close to, their specific growth rate there is a degree of instability in the system. Finally, if you have washout when an environmental factor is changed, you can only establish by educated guesses what might be limiting growth.

The proper functioning of a chemostat depends on continuous growth of all the biomass. That sounds a truism, but it is of particular importance when considering growing filamentous fungi in continuous culture. In batch liquid culture, we expect such fungi to grow as pellets. However, as they increase in size (see Section 5.3), the proportion of the mycelium no longer contributing to the biomass increases. Filamentous fungi can only properly be grown in a chemostat when the hyphae break as a result of the shear forces caused by the impeller mixing the medium. This means only certain fungi can be used with success, because it is not easy to obtain the necessary fragmentation of the hyphae. With filamentous fungi there is also the problem of growth of mycelium on the wall of the fermentation vessel. Remember that such wall growth is analogous to the growth of a pellet, in that some of the biomass is taken out of the influence of the medium and therefore does not grow at the same rate as the biomass in suspension. Under these conditions washout can occur.

6.3.2 Equipment

Figure 6.3 shows the fermentation vessel of the chemostat (1–2 l is an appropriate volume for most experimental studies of steady-state growth), together with the parts of the control system treated diagrammatically as distinct modules. These modules control the speed of the impeller, amount of foaming (addition of antifoam), pH (addition of acid and alkali), medium addition, temperature and aeration. All commercially available fermenters have all the components as shown but they will differ in detail. It is very important, before using a fermenter, that you spend as much time as necessary in getting to understand how it works and ensuring that you can disassemble and reassemble the fermenter vessel so that it works satisfactorily each time a run is made. Before using the fermenter as a chemostat, carry out some runs with the fermenter in batch mode, that is, with a suitable volume of medium in the vessel prior to inoculation and the ports for the entry and exit of medium blocked. In this mode, there is much less chance of contamination. Thus you can get used to the operation of the fermenter with worries about possible contamination reduced.

The fermentation vessel is sterilized by autoclaving with all the probes and relevant tubes in place for the entry of foam, acid/alkali, medium, air (exit as well as entry and, since there is no liquid in the vessel, the filters can be present on the two ports) and for the exit of organism and medium. The probes must be made of such a material as to be unaffected by autoclaving; make sure that this is so. Sterilization of filters is dealt with in Section 6.2. The medium feed line runs through a peristaltic pump which is part of the medium flow rate module (*Figure 6.3*) At the point where the feed line runs into the fermentation vessel a 'break' tube must be present within the line to prevent back contamination of the medium reservoir with cells from the culture (*Figure 6.4*).

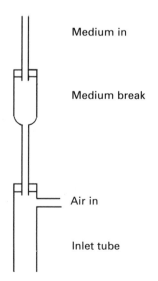

Medium in

Medium break

Air in

Inlet tube

FIGURE 6.4:   *Medium inlet tube assembly. The medium inlet port is kept free of organisms (which might otherwise reach it via an aerosol from the culture) by a stream of air. A second barrier to growth of organisms in the feed line is provided by the break tube.*

The rate of flow into the fermentation vessel is controlled by a peristaltic pump. The flow rate needs to be quantified (volume time^{-1}); this can be determined by the arrangement shown in *Figure 6.5*, which is part of the medium feed line between the medium reservoir (usually of 10–20 l capacity) and the peristaltic pump. The sampling port should be separate from the overflow tube. The walls of the latter can become covered with cells and, if the tube were to be used as sampling port, such cells could become loose and contribute to a sample, which would be unrepresentative. The sample port should have a tube of short length which should be flushed through before a sample is taken. Remember that when sampling the volume taken should be less than 10% of the culture volume so that steady-state conditions are not too disturbed.

pH is controlled by the addition of acid or alkali, control being via sensing by the pH electrode and monitored on a chart recorder. You must keep in mind that, in experiments concerned with the effect of various pH values on growth rate, the effect of changes in hydrogen ion concentration will be confounded by differing concentrations of anion, for example Cl$^-$ if HCl is used to lower the pH, and cation, for example Na$^+$ when NaOH is used to raise the pH. Organic buffering

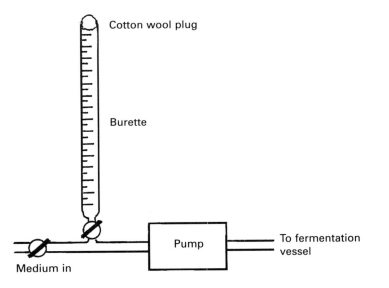

FIGURE 6.5: *Arrangement for determining the medium flow rate. The burette is filled from the medium reservoir via a three-way tap, which is then closed off. The tap is turned so the medium is fed into the fermentation vessel only from the burette. This allows the time for the descent of a set volume due to the action of the pump to be determined.*

agents can be used; they become expensive when used in this way. Check that any you use cannot be metabolized by your experimental micro-organism.

Aeration of the culture is achieved by sparging with air under constant pressure and by stirring. Air pressure is set by the use of gauges associated with the fermenter. Nevertheless, the air supply (gas cylinder) should be fitted with a constant pressure gauge to rule out the effect of fluctuations in pressure which can occur and cause difficulties. Aeration of the culture in this way should result in an oxygen supply more than enough to ensure that it is not limiting the growth of the culture. This is easily checked by the trace on the chart recorder which displays the dissolved oxygen concentration sensed by the oxygen electrode, when it can be observed that there is normally a considerable amount of the gas in the medium. The value for the limiting oxygen concentration can be determined by turning off the air supply and following the dissolved oxygen concentration as a function of time. The reader should consult ref. 2 for the background theory.

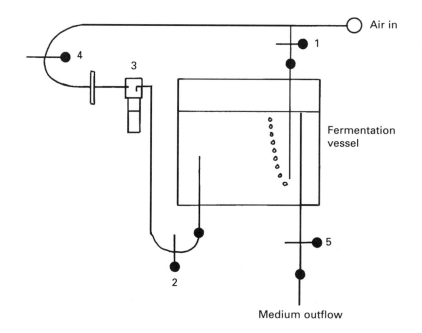

FIGURE 6.6: *The arrangement of taps/clips which need to be operated to ensure satifactory sampling of the fermentation vessel during continuous culture. This figure is required for use in conjunction with* Table 6.2.

The action of the impeller and the sparging of air will lead to foaming, the amount being dependent on the nature of the medium, the cells and extracellular products. Foaming is prevented by the addition of an antifoaming agent at a constant rate, determined by trial and error.

Wall growth can be a problem. A plastic-coated bar magnet, operated by another magnet outside the fermentation vessel, can be used to scrape the walls, though this must be done with the impeller stationary. The culture must be grown on for several hours before it can be considered to have returned to a steady-state.

Finally, there are the two matters of inoculation of the culture and the collection of samples. Inoculation can be performed with a syringe fitted with a needle which is used to pierce a seal, or by the addition of a larger volume of culture through a port of relatively wide diameter. The design of the port can allow for both possibilities. The virtue of adding a relatively large volume (50–100 ml) is that the

TABLE 6.2:   *The sequence of opening/closing of taps/clips which are necessary to obtain a sample from the fermentation vessel while the chemostat is in operation. The numbers of the taps/clips are those in* Figure 6.6

	Taps/clips					
Conditions	**1**	**2**	**3[a]**	**4**	**5**	**Comments**
Normal running	Open	Closed	Closed	Closed	Open	Air bubbles through medium and vents through outflow
Start sampling	Open	Open	Open	Closed	Closed	Pressure in vessel rises; culture flows through sample tube into bottle as air in bottle and filter is compressed
Stop sampling	Open	Open	Open	Closed	Open	Pressure in vessel drops; flow of culture into bottle stops
Clear sampling	Closed	Open	Closed	Open	Open	Air now enters vessel through filter and sampling hood, thereby cleaning line, and vents through outflow tube
Normal running	Open	Closed	Closed	Closed	Open	

[a] Sample bottle: closed = absent; open = present.

culture can start with a large inoculum which will have a good chance to grow. Only a relatively small number of replacements of the volume of the medium in the fermentation vessel are required to clear the culture medium in the vessel of the inoculum medium. Clearance can be checked by establishing when the composition of the outflow medium becomes constant, as determined by the concentration of one or two key constituents, for example glucose.

The medium is sampled by allowing it to feed, via the sampling port, into a sterile screw-topped bottle which can be attached by screwing into the hood over the port. A diagram of the relationship of the port to the fermentation vessel is shown in *Figure 6.6*. Collection of a sample involves the sequential opening of clips/taps as indicated in *Table 6.2*. The sequence outlined leads first to the filling of the sample bottle following the build-up of a positive pressure in the vessel and, secondly, at the end of that operation to the return of the chemostat to normal, without contamination of the medium in the vessel.

References

1. Tempest, D.W. (1970) in *Methods in Microbiology* (J.R. Norris and D.W. Ribbons, eds). Academic Press, London, Vol. 2, pp. 259–276.
2. Evans, C.G.T., Herbert, D. and Tempest, D.W. (1970) in *Methods in Microbiology* (J.R. Norris and D.W. Ribbons, eds). Academic Press, London, Vol. 2, pp. 277–327.
3. Drew, S.W. (1981) in *Manual of Methods for General Bacteriology* (P. Gerhardt, ed.). American Society for Microbiology, Washington, DC, pp. 151–178.

7 Isolation from the Natural Environment

7.1 Introduction

This chapter is not about microbial ecology: for information about how to proceed in this field of study, you need to consult relevant texts. Here we concentrate upon the procedures that should be used to isolate a particular microbe from its natural environment. The microbe of interest is known to be present in a particular habitat and it has to be identified; there is a need to know whether a known organism is present in an environment under study; a micro-organism representative of a particular environment has to be isolated for physiological, biochemical or genetic investigation.

It is, moreover, important to consider in detail any plans to isolate an organism which is thought to be representative of a study environment. In particular, two aspects demand attention. The first is whether the isolated organism is a true inhabitant rather than a casual visitor. The true representative is, in almost all instances, that which is able to complete its life cycle in the environment concerned. The only exceptions are certain specialized plant pathogens, whose life cycle involves two hosts. For a fungus, the capability to reproduce sexually (if that capability is present) in the environment is an additional but important criterion. The casual representative, which may be isolated from that environment, will be in the form of spores or nondividing cells coming from elsewhere. There can be a debate about those fungi that come into an environment from elsewhere via spores and produce extensive vegetative growth but never produce any reproductive structures, as to whether or not they are casuals. The second aspect concerns the meaning of the term 'environment'. For a large organism, such as a higher plant or a mammal, the environment can be defined relatively easily, because the size of the organism is such that one can use the characteristics of the ambient bulk features of the environment. Such an environment can be characterized by macroscopic probes. However, data obtained in this way may be of little help when considering individual microbes in nature. Their size enables them to live in sites within a particular habitat, for example soil, which has characteristics markedly different from the bulk features. Determining those particular characteristics presents a considerable challenge to the microbial ecologist. The corollary of what has just been said is that isolation of micro-organisms from a particular environment, seemingly incapable of living there on the basis of its bulk features, should not necessarily be ascribed to contamination. It goes without saying that isolation procedures should be designed not only for organisms likely to be found on the basis of bulk characteristics of the chosen study environment but for those that might not be thought *a priori* to be present.

7.2 Direct isolation

In view of the above, there is much to be said for direct examination of the habitat under investigation – either by bright-field microscopy or with the scanning electron microscope. With the latter, the fungal spores that can be identified, and the shape and nature of association of bacterial cells, can provide a guide to the types of organism that may be present as well as the type of material (substratum) with which they are associated.

Whether or not one has been able to identify any of the micro-organisms present, the need is to get them into culture, either collectively or individually. If you know what sort of organism you are trying to isolate, then use those recognized procedures and media which select for the kind of organism that interests you (Chapter 3). Solid media, if they can be used, are by far the best for isolating micro-organisms, because of the formation of discrete colonies (remember, if you do have initial difficulties in obtaining discrete colonies, they can always be obtained by dilution of the inoculum). Under these circumstances, colony morphology is readily observed and this characteristic aids identification, along with those other chacteristics which are readily observable following microscopic examination of parts of the colony picked from the culture surface.

Tolerance to extremes, for example high temperature or salinity, aids isolation. When conditions are less stringent, isolation becomes more difficult, since many more micro-organisms are adapted for growth under conditions of average temperature, water and nutrient availability. As will be seen, the central issue to be faced is the isolation of a particular species of micro-organism from those others that may be present.

It is important to always keep in mind that once the process of isolating micro-organisms from the habitat of interest has been initiated, potentially pathogenic organisms may be isolated. There is no way this can be avoided completely, but it can be minimized by the avoidance of media that favor such organisms. Of course, the chances of isolating pathogenic organisms are greater with some environments than others. There is little doubt of the high probability of doing so if one were studying soil contaminated with sewage waste. Samples from such an environment must not be examined without the necessary facilities (see Section 2.3). But you may not be properly aware of the previous history of an environment, so you need to be vigilant.

There are two ways of encouraging the growth of micro-organisms on the solid medium of your choice. Small fragments of the solid habitat or a small volume of the medium from an aquatic habitat can be spread over the growth medium. The other way is to shake the solid medium with sterile distilled water, then spread a small volume of the resulting aqueous suspension over the growth medium. The advantage of this latter procedure is that it dislodges microbial propagules from the material and dilutes them (and there can be still further dilution before application to the growth medium), so that the chance of one colony occupying the same space as another is reduced. Also, the aqueous suspension can be passed through filters of known porosity to separate bacteria from fungal cells and propagules.

It may be necessary, when dealing with solid material, to make the isolation procedure a two-stage process. The material is first washed in sterile distilled water (and, if it is thought desirable, small volumes of the washing liquid transferred to solid media to examine the micro-organisms contained therein). Then small fragments of the washed material are themselves placed on the solid medium. This two-stage process has been used with plant roots. Washing removes surface bacteria and fungal spores, and when the washed root is placed on agar only those organisms closely associated with the plant organ grow out on to the agar.

Finally, it may be necessary when searching for micro-organisms that may be residing within solid material, to surface sterilize that material to remove surface-associated organisms. This is especially true of living material such as plant leaves. It is customary to immerse the material for 2–10 min in sodium hypochlorite, usually in proprietary form, such as Domestos (Lever Bros) usually at 2.5% (v/v) – more recalcitrant material may require 5%.

7.3 Baiting

There is a long tradition in mycology of using baits to capture fungi from particular environments. Baits can be various, for example uncontaminated wood has been used for the capture of wood-decaying fungi in the sea; fruit to capture freshwater aquatic fungi and pine pollen to capture a particular class of marine fungi, the Thraustochytrids. Baits are placed in the appropriate habitat and removed after a suitable time. Often, the bait must then be incubated in an otherwise sterile moist environment and examined from time to time for the appearance of the desired micro-organisms. It may be necessary to wash the surface of the bait before such incubation to remove both casual organisms and detritus. Such washing is obligatory with wood baits, if they are left in the chosen habitat for any length of time. It is usually necessary to leave wood in a particular environment for long periods – up to several months – because there is a succession to the fungal colonization, some species not appearing until significant decay has taken place. A similar situation can hold for other baits. Incubation of a bait, once it has been removed from the habitat, may need to be of some duration, particularly for fungi that do not produce reproductive structures readily. Without such structures, most fungi cannot be identified.

8 Identification of Microbes

8.1 Principles involved in identifying microbes

Hundreds of thousands of different microbes have been meticulously described and more are still being discovered. A great deal of intellectual effort and much research energy has been spent organizing the information accrued and grouping together organisms with similar features. An organism can be isolated from the environment and its characteristics compared with those of others. A microbe can thus be assigned to a particular genus and species (*identification*). If the organism cannot be so assigned, its characteristics can be compared with previous records, its relationships with other organisms established and the unknown organism named.

Over many years rational schemes for distinguishing between microbes and for naming them (*nomenclature*) have been devised. Conventional schemes include information relating mainly to morphological characteristics supplemented with some physiological information (e.g. nutritional requirements, fermentation products, pigment production). Such characters define the *phenotype* of an organism; the physical manifestation of those parts of its genetic potential (*genotype*) that are expressed under particular conditions. The immense diversity of organisms is the result of genetic changes occurring due to recombination and mutation. Recently, however, data from physiological experiments, chemical analyses and molecular techniques have been included in classification schemes, making them more and more comprehensive and informative. Comparisons of nucleic acid chemistry, protein chemistry and gene sequences have increased our understanding of the ancestry of organisms and the evolutionary significance of relationships between organisms (*phylogeny*). It has been shown that phenotypic characters alone can be poor predictors of phylogeny.

Names and descriptions of newly described organisms are published in printed articles in widely available journals or books according to convention. The authors of that publication are then the *authorities* for the scientific names, and abbreviations of those names are used in references to the organism for clarity, for example *Septoria apiicola* Speg. (Spegazinni is the authority for this species of fungus). In some cases the name, or the rank, of an organism may be changed, usually in the light of new information. In that case the name of the first authority is given in parentheses after the species name and that of the author making the change follows outside the brackets, for example *Crinipellis perniciosa* (Stahel) Singer.

Type strains for bacterial species are published in *Approved lists of Bacterial Names* [1] and updates in the *International Journal of Systematic Bacteriology* [2]. Helpful information concerning identification and classification is given in *Bergey's Manual of Determinative Bacteriology* [3]. In order to prevent the publication of confusing or ambiguous names for fungi, conventions are followed as given in the *International Code of Botanical Nomenclature*. Changes to this code are debated and controlled by the International Botanical Congress (see ref. 4 for further information).

8.2 Microscopy

All aspects of microbiology involve the use of microscopy. Once a culture has been set up it can be examined by eye and physiological tests can be carried out on samples of the culture, but a significant amount of information can also be obtained by determining microscopically the size, shape and cellular growth habit, both qualitatively and quantitatively. Microscopy also allows the study of the patterns of differentiation which occur throughout growth and development on different substrates, and also reactions of cells to added chemicals and to the presence of other microbes. Light microscopy can distinguish between (*resolve*) particles as small as 0.2 µm in diameter and should be used routinely and frequently for checking cultures. Details of surface structures and internal organization of cells can be obtained by higher magnification using electron microscopy. This technique can resolve particles of 0.002 µm diameter. Both scanning electron microscopy (SEM), to give a view of overall cell morphology and cell surfaces, and transmission electron microscopy (TEM), to view thin sections through specimens, can be used to great advantage with micro-organisms, to obtain information about structure and development. TEM is essential for the investigation of subcellular structure.

8.2.1 Light microscopy

Often a working knowledge of light microscopes is assumed, but unless the instrument is set up and used correctly it will be difficult, if not impossible, to obtain good results, and information may easily be missed. Research microscopes may have many attachments and facilities so that specific local instructions will be required before any attempt to use these instruments. However, a simple microscope can give an excellent view of specimens, provided that it is set up appropriately. Remember that the optics of microscopes have provided extremely good resolution for very many decades; on newer models mainly the attachments and photographic facilities have been improved. Remember, too, that there is no substitute for practice, the more familiar you are with the equipment the more confidently you will be able to proceed. Additionally, the more familiar you are with the organisms in which you are interested the more critical you can be and the more rapidly you will spot any changes in the cultures and in the cells. Stereo, bright-field and phase-contrast are the most commonly available and widely used facilities that can be used to enhance the role of the simple light microscope. It is most important not to assume that you will be able to use any microscope with which you are faced. Consult a textbook for information [5] and take local advice so that you make best use of the available equipment.

Stereo microscopy. For some work in a microbiology laboratory the use of a stereo (or dissecting) microscope is ideal for examining the surfaces or margins of colonies (particularly fungal) growing on the surface of agar or on natural substrates. Such microscopes, though compound, are most akin to simple models, that is, a magnifying glass, giving magnification up to 100x. Modern instruments make use of either illumination from above the specimen shining on to the surface, or transmitted light passing up through the specimen. Stereo microscopes give a three-dimensional view of the specimen. Since these microscopes have a long focal length, Petri dishes can easily be placed under them for direct observation.

Bright-field microscopy. If a higher magnification is required, for inspection of individual cells or mycelium in liquid culture for example, then the use of a high-power compound or bright-field microscope will be required. Cells absorb and scatter light to different degrees, which renders them visible in the light beam through the lenses of the microscope. Using this type of instrument, objects can be seen by virtue of their translucence and contrast with the background. In some cases, especially if cells are particularly small and transparent, stains can be used to increase the contrast of the specimen (see Section 8.3.2).

The steps in the procedure for the use of a bright-field microscope are listed as follows.

- After plugging in, the illumination is switched on.
- The condenser is raised to the highest position.
- The low-power (x10) objective is placed into the light beam.
- The specimen is placed on the microscope stage and moved into the light beam.
- The specimen is brought into focus using the coarse and then the fine focus controls.
- Next the specimen is ignored and the *condenser* is focused, to give an image of the field lens (granular surface) in the eyepiece (this is known as critical illumination).
- The eyepiece is (CAREFULLY) removed and the condenser iris diaphragm closed so that it can just be seen through the eyetube. The iris diaphragm is adjusted until 70% of the aperture diameter is illuminated. The eyepiece is then replaced.
- Good focus and illumination should now be set for the x10 objective. The x40 objective lens can be swung in and focused (you will need to adjust the condenser iris as described above).
- The x100 objective can subsequently be swung in. Very little focus adjustment should be necessary if the above procedure has been followed. To give good resolution with this objective a small drop of immersion oil must be placed on top of the specimen and the objective lens swung round into the oil. The condenser iris should usually be more fully opened when using this objective. (At this magnification there is very little distance from the objective lens to the specimen, if this procedure is followed only minor adjustment to the focus will be necessary. If the x100 objective were to be swung in *without* focusing at the lower powers first, it would be extremely difficult to focus on the specimen.)
- The objective lens should always be swung in from the same (clockwise) direction, so that wear on the positioning groove is minimized.
- It is *essential* to wipe the objective lens free from immersion oil. The microscope should be left with the x10 objective in position and the whole instrument must ALWAYS be covered when it is not in use.

A very great deal of useful information can be gained using a well-set-up bright-field microscope. In many instances no further detailed microscopy may be required.

Phase-contrast microscopy. Stains must be used to increase contrast for bright-field microscopy, which usually results in the death of the specimen. The phase-contrast microscope works on the principle that light travels through different materials at different rates. Cells have a different refractive index from the surrounding medium and produce a phase difference in the light that travels through them. That phase difference is used to create greater contrast in the specimen and a clear view of otherwise poorly contrasting material can be seen. Dense materials, such as nuclei, appear dark, whereas less dense materials, such as cytoplasm, appear bright. This is a particularly useful system for resolving small cells and determining the presence or absence of cellular contents.

Fluorescence microscopy. This technique is used to increase the contrast of specimens by staining with specific dyes (fluorochromes), which absorb energy from short light waves and then emit light of a longer wavelength. Labeled molecules can then be identified in specimens, for example DNA can be detected under UV fluorescence by the use of the DNA-specific dye 4′,6-diamidino-2-phenylindole (DAPI) which is particularly useful with fungi, since fungal nuclei are extremely small and difficult to detect. Fluorescence microscopy is often combined with confocal microscopy by means of which thick tissue sections labeled with fluorecent dye may be examined and a three-dimensional view of the specimen obtained.

Nomarski interference microscopy. This technique makes use of differential interference contrast to increase contrast in living cells and creates an image which appears three-dimensional (see ref. 5). It has proved particularly useful in tracking the movement of organelles inside cells since living tissues can be viewed.

Dark-field microscopy. Dark-field illumination makes specimens appear bright against a dark background. The microscope is set up with a specialized condenser so that only the light hitting the specimen passes through the objective lens. This technique can be used for specimens with poor contrast and is most useful for those organisms with distinctive morphology. Both stereo and compound microscopes can be set up for this dark-field illumination.

8.2.2 Electron microscopy

This technique is vital for many areas of microbiology where high magnification and high resolution are required. The results of electron microscopic investigations are usually recorded by photography. Exposure to electrons alters the appearance of the specimen and it is usual to take photographs fairly quickly to avoid any damage to structures in the electron beam. Electron microscopy is highly specialized, requiring some experience both in specimen preparation and in operation of the instruments to give successful results. Expert advise, relevant to the best method of preparing the specimen under consideration and the particular instrument to be used, must be sought; most electron microscopes are usually operated by trained personnel.

Transmission electron microscopy. Transmission electron microscopy (TEM) is used for the investigation of the internal structure of cells. After fixing and dehydration, ultrathin sections of the specimen are taken and adhered to a tiny copper grid for viewing. The grid is placed into the microscope chamber where a vacuum is created and a beam of electrons, focused by electromagnets, is passed through it. The scattered electrons then form an image on a fluorescent screen. This technique can be used to resolve particles 0.001 μm apart. Ultrathin sections are required and usually special stains are used containing heavy metals (e.g. osmium tetroxide, uranyl acetate, lanthanum or lead citrate, compounds containing elements with high atomic weight which therefore scatter electrons well) to enhance contrast by increasing the electron density of some cellular components.

Scanning electron microscopy. Scanning electron microscopy (SEM) allows magnification of the surfaces of specimens and the investigation of external features only. The preparative techniques used are generally less time consuming than for TEM. Specimens, which may be quite large (0.5–1 cm blocks), are dehydrated, stuck on to a support and vacuum-coated with metal, for example gold or platinum. The prepared specimen is then placed in the microscope and an electron beam scans over the surface. In this way a three-dimensional image of the specimen is obtained with a good depth of field, and magnifications between x15 and x100 000 can be achieved. Surface structures of microbes can be seen, which is particularly useful for the examination of fungal spores, and the relationship between microbes on the surface of a particular substrate can be identified. It is also possible to detect microbes growing within tissues (e.g. higher plant stems, leaves and roots) by viewing the surface of a cut face of the tissue. SEM, though seemingly more restrictive than

TEM in that it provides no information about internal organization, is nevertheless a more powerful tool in microbial taxonomy and the study of the organization of colonial microbes, such as actinomycetes and fungi.

8.3 Preparation for light microscopy

It may be necessary to observe specimens under the microscope in several different ways, depending on the nature of the available culture and the information sought. In general, techniques can be considered in two groups, those for use with live cultures or wet material and those where a dried, stained sample is used.

8.3.1 Wet-mounting and examination of live cultures

A *wet mount* is the most straightforward procedure and a useful means to start initial investigations of a culture. A small sample is placed on to a microscope slide and covered with a coverslip. The sample may need to be mixed with a little water on the slide to wet spores or to spread cells for observation under the microscope. A *hanging drop* preparation can be made in a similar way, using a slide with a small depression in the centre (cavity slide). A drop of suspended culture is placed on to a glass coverslip and inverted so that the drop hangs over the depression. The coverslip may alternatively be raised by a ring of Plasticine (*Figure 8.1*) and used with a flat slide. This is a particularly useful method for the

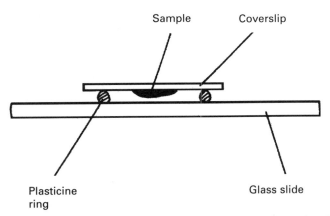

Sample   Coverslip

Plasticine   Glass slide
ring

FIGURE 8.1: *Hanging drop method for wet-mounting live cultures.*

investigation of cell motility since movement is not restricted by the weight of a cover slip. It may be easier to see motile cells by focusing on individuals near the edge of the drop rather than in the center where cells may move in and out of focus rapidly.

In order to examine fungal colonies for the presence of spores, it may be possible to view the colony directly on the surface of agar using a stereo microscope. The aim is to view aerial mycelium undisturbed so that natural spore arrangements are not dislodged. A small piece (1–1.5 cm^2) of fresh transparent sticky tape may be used to pick up spores and aerial mycelium from the surface of a culture. Using flamed forceps, the tape must be pressed firmly on to the colony (sticky side down) and then lifted directly upwards, away from the culture. The tape can then be stuck on to a microscope slide and viewed without a coverslip (the tape acts as a barrier). Sporulating structures can then be viewed in their natural form. This is a very useful and successful method, but care must be taken not to release spores into the atmosphere. Slides, plus tape, must be disposed of into disinfectant immediately after viewing. This method is recommended only for cultures in containment classes I and II (see Section 2.2).

8.3.2 Staining techniques

Micro-organisms are often small, transparent and contain little pigment. As a result they are not easy to see under the microscope. Staining techniques can be used to increase the contrast of the organism against the background. Additionally, some specific microbiological stains are used to distinguish particular materials within cells. Some stains can be purchased ready to use, whereas others may need to be made up freshly in the laboratory. The detailed composition of individual stains is largely beyond the scope of this text, although some of the more commonly used stains have been itemized (consult ref. 6 for further information). In some instances, a number of alternative methods are available. Those given here are widely accepted and have given good results for the authors. However, if these methods are not successful for you with your material, we recommend you try alternative protocols [6].

Bacteria.
Gram staining. This technique is used to distinguish between types of bacterial cells by virtue of fundamental differences in the structure of cell walls. This procedure is one of the most commonly used techniques in microbiology laboratories. Gloves should be used for protection from the dyes.

- Using aseptic techniques and an inoculating loop, a small portion of a bacterial colony is placed on a clean, grease-free slide (slides are preferably handled using forceps throughout) and mixed with a little water, spreading out the cells from the colony across the surface of the slide. This gives a thin smear of bacteria. It is essential not to have too many cells in the smear or it will be too dense to view successfully.
- The smear is allowed to dry in air and then the slide is passed (face upwards) two or three times through a Bunsen flame. This will 'heat-fix' the cells to the slide. Care must be taken not to heat the slide too much or the cells will become distorted.
- The slide is then ready for a series of staining and washing steps. Staining is usually carried out on a rack, above a sink with running water near by.
- The heat-fixed smear is flooded with crystal violet dye (0.5% w/v aqueous solution) for 30 sec and washed with water.
- The slide is flooded with Lugol's iodine (a mordant used to fix the dye inside the cells) for 1 min and then washed with water.
- The slide is then rapidly washed with 95% (w/v) ethyl alcohol (2–3 sec) and subsequently rinsed well with water. This step will decolorize Gram-negative cells; Gram-positive cells will retain the purple dye.
- An aqueous safranine solution is used to counterstain Gram-negative cells. The slide is flooded with safranine solution (0.5% w/v) for 30 sec, then washed with water and blotted dry.
- When the slide is dry, the stained cells are observed under the microscope using the procedure described in Section 8.2.1. It is not necessary to use a coverslip, after focusing with x10 and x40 objective lenses the immersion oil is placed directly on to the smear and the oil-immersion objective (x100) can be swung in.
- Gram-positive bacteria retain the crystal violet dye and appear PURPLE. Gram-negative bacteria are washed free of the dye and counterstained with safranine so that they appear PINK.

Although this technique is very useful, some species can show different reactions (Gram variable) at different stages of growth. It is important to standardize the technique where comparisons are to be made, and it is useful to include a sample with a verified Gram stain reaction along with any organisms of unknown Gram reaction. *Escherichia coli* cultures are usually used as an example of Gram-negative cells (staining pink). A smear of *E. coli* cells may be placed on one end of a clearly marked slide and an unknown organism smeared on the other end, allowing direct comparisons to be made easily after staining. Endospores are not stained by this treatment.

Flagella stain. Flagellae are so small that they are difficult to resolve under the light microscope. In order to see them they must be treated with a mordant which increases their apparent diameter and makes them more visible. Also they are easily dislodged from the cells, so this technique must be carried out carefully. A small drop of culture containing motile cells is placed on to one end of a clean grease-free slide, allowed to run along the length of the slide and then air dried. The dry smear is flooded with picric acid–tannic acid mordant for 5 min, washed well and drained. The smear is then treated with hot, sensitized silver nitrate solution for 5 min, washed and blotted dry with tissue. Flagellae will appear as brown threads. It may be helpful if the amount of light passing through the specimen is reduced by closing the iris diaphragm, in order to see the flagellae more easily.

Acid-fast staining. Some bacteria, known as 'acid-fast' bacteria, cannot be easily decolorized with acid–alcohol after staining with carbolfuchsin stain. This is a property used as a diagnostic test to distinguish actinomycetes and some mycobacteria.

- A film of bacteria is smeared on to a glass slide, air dried and heat fixed.
- The slide is then flooded with carbolfuchsin (basic fuchsin, 0.3 g; ethanol 95% (v/v), 10 ml; heat-melted phenol crystals, 5 ml; distilled water, 95 ml) and heated over a Bunsen burner flame or hot plate. Heating is continued until steam rises for 5 min (it is important not to boil the stain).
- The slide is washed under tap water until no more color is removed in the washing water.
- The bacterial smear is subsequently flooded with decolorizing agent (acid ethanol–ethanol 95% (v/v), 97 ml; concentrated hydro-chloric acid, 3 ml) and then washed immediately. This step is repeated with fresh decolorizer until the smear has only faint pink coloration.
- The smear is then counterstained with methylene blue chloride (0.3 g 100 ml^{-1}) for 20–30 sec, washed well with tap water, blotted dry with blotting paper and examined using a microscope.
- Acid-fast bacteria will appear RED and nonacid-fast bacteria will appear BLUE.

Endospore stain. Some bacterial species form endospores within their cells. These structures have importance, because they are highly resistant structures and can remain dormant for very long periods of time. Endospores will survive high temperatures, desiccation and chemical agents that cause disruption to vegetative cells. A smear of cells is stained with 5% (w/v) aqueous malachite green for 5 min, heating the slide carefully over a gentle flame until it steams. It is then washed with water for 30 sec and counterstained with 0.25%

(w/v) aqueous safranine for 30 sec, washed with water again and blotted dry. Spores will be stained green and the rest of the cells will be red.

Examination of fungi. For investigations of fungi, it is usual to place a small sample of fungal tissue on to a clean glass slide, add a few drops of water and a coverslip, to enable examination under the microscope. Lactophenol is often used as an alternative mounting medium to water because it is slightly viscous, clinging around hyphae. Lactophenol can be purchased ready to use or can be made in the laboratory as follows: phenol crystals, 20 g; water, 20 ml; lactic acid, 16 ml; glycerol, 31 ml). Cotton blue dye (aniline blue) may be added (0.05 g to the above quantity) to increase the contrast around hyphae and other structures. The dye does not penetrate into the hyphae as in the case of bacterial stains. In our experience, this method works best if the lactophenol mixture is used sparingly.

8.3.3 Measurement of organism size

The size of a microbe, or particular structures and appendages, can be determined using a microscope and a pair of micrometers. The eyepiece micrometer rests on a ledge inside the microscope eyepiece and the scale on it is therefore superimposed on the image of the specimen when viewed through the eyepiece. The dimension of the microbe can then easily be recorded in eyepiece units. The actual value for each eyepiece unit is not known in absolute terms but can be calibrated using a stage micrometer. These most usually have divisions of 100 µm. When this is on the microscope stage the accurate scale will appear with the eyepiece micrometer scale superimposed on it (*Figure 8.2*). It is necessary to line up the two scales, by moving the stage micrometer, until a whole number of eyepiece units coincide with stage micrometer divisions. In the example shown, four eyepiece units are equal to one slide division, that is 100 µm. Therefore:

$$1 \text{ eyepiece unit} = \frac{100}{4} = 25 \text{ µm}.$$

For a specimen with a diameter of 0.2 eyepiece units:

$$\text{Actual diameter} = 25 \times 0.2 = 5 \text{ µm}.$$

The eyepiece micrometer must be calibrated for each microscope objective.

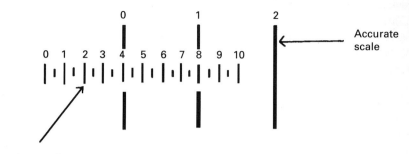

Eyepiece scale

FIGURE 8.2: *Calibration of an eyepiece micrometer.*

8.4 Diagnostic tests for bacterial identification

Initial steps in the identification of bacteria will be concerned with the establishment of a pure culture, the description of colony and cell morphology and the results of diagnostic staining (e.g. Gram stain, acid-fast staining, see Section 8.3.2). Further tests are required based on growth and activity on selective media. It is beyond the scope of this text to cover all such tests but the principles of some key procedures and those most commonly used are considered below.

It is sensible to adopt a reasoned approach to testing cultures, and guidance can be obtained from published keys for particular bacterial genera and families. Diagnostic tests often make use of chemical changes in the medium, brought about by growth of the inoculated organism, to elucidate the properties and effects of that organism. Indicator compounds, changing color with fluctuation in pH, and other indicators of chemical alterations are often incorporated into the media used for diagnostic tests.

Aerobic/anaerobic growth. The presence of obligate anaerobes is easy to miss in conventional aerobic culturing. These microbes require special handling to ensure that contact with oxygen is avoided (see Section 4.2.3). However, the distinction between aerobic and anaerobic growth may not always be clear. Some species are able to grow in the presence or absence of oxygen. Changing the redox (Eh) potential of the medium will provide a means for investigating anaerobiosis. *Redox potential* is a measure of the ease with which electrons are lost (substances oxidized) or gained (substances reduced). Redox dyes (reversibly oxidized and reduced) can be used to alter the redox potential in a culture medium and will indicate change in redox by a change in color. Reducing agents (e.g. sodium thioglycolate) can also be added to culture media to depress redox levels.

Oxidation/fermentation of sugars. To establish whether carbohydrate is utilized by oxidation or fermentation, Hugh and Leifson's basal medium (which includes bromthymol blue as a pH indicator) containing a test carbohydrate is made up in test tubes (see ref. 2 for working details). These are heated in a boiling water bath to remove most of the dissolved oxygen, cooled quickly and then two tubes are stabbed with inoculum using a straight wire. The medium in one tube is then overlaid with sterile liquid paraffin (mineral oil) (anaerobic tube), the second tube is left without an overlay. Both tubes are incubated under appropriate conditions for growth. Oxidizers show acid production, turning the indicator from blue to yellow, in the open tube only. Fermentation is shown by acid production in the covered tube and in the open tube (e.g. *Pseudomonas* – aerobic acid production; enteric bacteria – anaerobic acid production).

Catalase test. The presence of active catalase enzyme results in the release of bubbles of oxygen gas from hydrogen peroxide (H_2O_2). Grow up an active culture of the test organism, scrape off some cells with a glass rod and mix with 3% hydrogen peroxide on a clean glass slide. After a few minutes inspect for the presence of bubbles of oxygen (*Staphylococcus* +; *Streptococcus* –; *Bacillus* +; *Clostridium* –).

Oxidase test. Grow up colonies on agar-based nutrient medium. Flood the plate with reagent (1% α-naphthol in 95% ethanol and 1% aqueous tetramethyl *p*-phenylenediamine oxalate in equal volumes). After a few seconds the presence of blue or brown colonies indicates a positive reaction (*Pseudomonas* +; enteric bacteria –).

Nitrate reduction. Inoculate a liquid medium containing nitrate and incubate at a temperature conducive to growth (in some instances a semisolid culture is used to promote semianaerobic conditions – the development of gas bubbles is an indication of denitrification). Controls lacking nitrite are set up simultaneously. The presence of nitrite is then detected with α-naphthylamine-sulfanilic acid reagent and a red coloration indicates a positive test (*Pseudomonas*, denitrification; *Escherichia coli,* nitrate to nitrite).

Casein hydrolysis. Autoclaved skimmed milk combined with nutrient agar is inoculated and incubated under appropriate conditions for growth. A clear zone around the colony indicates hydrolysis (distinguishes between *Bacillus* species).

Starch hydrolysis. A medium containing soluble starch (lacking glucose) is inoculated and incubated. After growth the plate culture is flooded with iodine solution (Lugol's iodine used for Gram staining) and positive tests for starch hydrolysis are indicated by the presence of clear zones around the colonies. Unused starch stains dark purple.

Hydrogen sulfide production. Agar medium containing iron citrate (ferric ammonium citrate 0.025%, or ferrous sulfate 0.015%), peptone (source of organic sulfur) and sodium thiosulfate (up to 0.3%) is inoculated, usually as a stab culture (to provide anerobic conditions). If H_2S is produced during incubation, a black precipitate will be seen along the line of inoculation (*Proteus vulgaris* +; *Escherichia coli* −). More sensitive methods are available (see ref. 6).

Tests for identification of the Enterobacteriaceae. These tests might be carried out if fecal contamination is suspected, and will aid in the identification of the Enterobacteriaceae (see ref. 6).

Indole test. This procedure tests the production of tryptophan from proteins metabolized to indole. A medium rich in tryptophan (peptone water) is used to grow up bacterial cells. The culture is tested by adding 1 ml xylene and agitating vigorously. Subsequently 0.5 ml dimethyl aminobenzaldehyde (Ehrlich's reagent) is added by pouring down the side of the tube to form a layer between the broth culture and the xylene. A positive test is indicated by the presence of a red-colored ring below the xylene (*Escherichia* +; *Enterobacter* −).

Methyl red test. Organisms carrying out mixed acid fermentations produce acid. Inoculated glucose-peptone medium containing methyl red as an indicator (yellow at pH 4.5 or higher, and red at lower pH) will show a red color where mixed acid fermentation has occurred (*Escherichia* +; *Enterobacter* −).

Voges–Proskauer test. Inoculate two tubes of methyl red Voges–Proskauer broth, containing polypeptone, and incubate one at 37°C and the other at 25°C for 48 h. The second stage tests for the presence of acetoin using α-naphthol. A 1 ml aliquot is extracted and mixed with 0.6 ml 5% (w/v) α-naphthol in absolute ethanol. The reaction depends on the presence of oxygen at this stage and tubes are usually left at an angle, to increase the surface area : volume ratio, for 15–60 min. A positive test is indicated by the development of a red coloration in the mixture (*Enterobacter* +; *Escherichia* −).

Citrate utilization. The utilization of citrate as sole carbon source is indicated by the alkalinization of the growth medium. Citrate medium is made up including bromthymol blue as a pH indicator. At alkaline pH the indicator is blue and the test is positive (*Enterobacter* +; *Escherichia* −).

A series of such growth-dependent *diagnostic tests* is used to place unknown bacteria into particular taxonomic groups. Many such tests are carried out daily in hospitals and pathology laboratories, where they are extremely important in accurate diagnoses. As a result of the need for rapid, reliable and reproducible results a number of commercial kits are now available [7] to enable these tests to be performed in a uniform manner and to yield reliable results routinely. For example, the API 20E system can be used for distinguishing between members of the Enterobacteriaceae in place of the specific tests mentioned above. Such kits usually comprise small-scale aliquots of diagnostic media contained in tiny wells, allowing a number of tests to be run simultaneously. A small plastic strip of wells each containing different, ready-prepared selective medium, is supplied. Each well must be inoculated and the whole strip incubated according to the manufacturer's instructions. Since many of the tests rely on color changes in the medium, manufacturers also supply color charts to allow assessment of positive and negative results. From this the most likely identification of the organism can be made, and the tests repeated if necessary.

8.5 Numerical taxonomy

The term 'numerical taxonomy' refers to arrangements of taxonomic groupings (key generation) produced by statistical assessments of similarity between characteristics. Such systems have been used widely and effectively for organizing groups of bacteria, but has been used much less for grouping fungi, algae or protozoa.

Numerical taxonomy makes use of as many characters as possible to determine similarities between the organisms included in tests (Operational Taxonomic Units) to define the taxa. All the characters are weighted equally, a matter that has caused a great deal of controversy amongst taxonomists [8]. It may be argued that some characters are more important than others in determining taxa. However, in constructing numerical classifications all characters are given equal importance. Since large numbers of characters are usually included, numerical taxonomy makes use of carefully designed computer programs to handle the analysis of the large amounts of data. The stages involved in designing a numerical system for bacterial taxonomy are shown in *Table 8.1*. A few general points concerning the ways in which numerical taxonomy schemes are devised will be considered here, but for more detailed information the reader should consult ref. 8.

Large numbers of strains (±600) can be included in devising a numerical taxonomy scheme, and since the data will eventually be handled by a computer system, there is little reason not to include a reasonably large number. Included with any new isolates or unknown strains should be organisms of known identity and, in particular, well-established isolates from culture collections to provide reference strains (see Appendix B). It is essential to use only pure cultures, otherwise erroneous results may be obtained which might influence the final definition of the categories. Duplicate strains should also be included as a test for the reliability of the system.

In selecting tests to be included it is advisable to use as many as possible (100–200) and to ensure that the selection represents as much of the biology of the organisms concerned as possible. Since a large number will be included, it is useful to select tests that can be performed most easily and to standardize the procedure for each test carefully. Additionally, tests that are not easily reproducible should be

TABLE 8.1: *Numerical taxonomy of bacteria*[a]

• Strain selection
      Pure cultures required
      Include reference cultures
      Include replicates

• Test selection
      Large number of rapid tests required

• Record results
      Analyze test error

• Code data

• Computer analysis
      Calculate similarities
      Cluster analysis

• Interpret results
      Define clusters
      Produce identification system
      Select representative strains

[a] Simplified from ref. 8 (p. 15, Fig. 2.1) with permission from Chapman & Hall.

avoided (although reproducibility is not always easy to ensure). Many morphological characters can be included as well as physiological and biochemical tests, for example colony morphology, cell shape, motility, nutrient requirements, fermentative metabolism, presence of specific enzymes, nitrate reduction, utilization of compounds as sole sources of carbon. Use of commercial testing kits (Section 8.4) can increase the speed and accuracy with which tests can be carried out. Eventually, test results are coded positive or negative and entered into a computer program for analysis of the similarities between the organisms included in the tests. It is often difficult to determine if a test is indicating a negative or positive reaction. Clear guidelines must be decided and adhered to. The inclusion of a number of replicate treatments may help to determine the reproducibility of individual tests.

It is then possible to order the organisms into groups with high similarity. Hierarchical methods produce a classification which is ranked, grouping strains into species, genera and families. Strains are grouped together (cluster analysis) according to the degree of similarity between them and a similarity matrix or a dendrogram may be drawn [8]. Most groups of bacteria have now been well studied and classification systems have been published, so that it is relatively easy to establish how well a new isolate fits into an accepted scheme by the method of numerical taxonomy.

8.6 Chemotaxonomy

Since the advent of highly sensitive and very reliable analytical techniques there has been an increase in the use of molecular markers for taxonomic purposes, to great effect. Chemical and molecular methods are now used to augment and, in some instances, strengthen the more morphological and physiological approaches taken previously. Detailed protocols will not be given here but the salient points will be considered.

The growth of microbes under different environmental and nutritional conditions may have a marked influence on the relative chemical composition of those cells that are produced. Many of the chemical components of cells vary markedly with the environmental conditions under which they have grown, for example temperature affects the lipid composition of membranes. Additionally, the relative composition of cell walls may be influenced by the availability of nutrients in the environment during growth and also, sometimes, by the stage of development. It is obviously essential, therefore, that for chemosystematic studies the cultural conditions are well defined and precisely adhered to. However, although the amounts of nucleic acids within an organism vary with changing growth rates, the composition of this DNA and RNA does not alter. Comparisons between these components are therefore extremely useful for taxonomic purposes.

Analysis of DNA base composition. Comparisons between the DNA base ratios of bacteria have shown that bonding occurs between the nucleotide bases guanine (G) and cytosine (C) and also between adenine (A) and thymine (T), but the relative amounts of the two base pair combinations varies between different species. The ratio is normally expressed as the G+C ratio (mol% GC) and this varies from just under 25% to just over 75% for prokaryotes, 25–60% for fungi and 38–70% for algae, and is constant for a particular species. However, this is a negative guide. Two bacteria with different GC ratios can be said to be unrelated. If two bacteria have a similar GC ratio they may be related, but not necessarily. However, if they also have similar DNA sequences then they are likely to be related.

Nucleic acid hybridization. The degree to which DNA sequences are similar (sequence homology) will determine the degree to which those bacteria are related. This can be determined by assessment of the degree of base pairing (nucleic acid hybridization) between DNA molecules from the two species. Such experiments must be carried out using carefully prepared DNA from the cells in question. Cells are initially lysed with detergent, protein is removed with a nonspecific protease and treatment with chloroform. RNA can be removed by ribonuclease (RNase) digestion and DNA precipitated in isopropanol. Determination of the nucleotide composition is carried out after DNA hydrolysis and quantified by high performance liquid chromatography (HPLC). DNA homology experiments can be used most usefully to compare closely related organisms. More distantly related organisms can be usefully compared by the degree of homology between DNA from one organism and rRNA from another: DNA : rRNA hybridization.

Ribosomal RNA (rRNA) molecules are essential for protein synthesis in all living organisms and have a distinctive nucleotide sequence in each species. Therefore, segments of rRNA from different organisms can be sequenced (particularly 16S rRNA) using relatively crude cell extracts. Those sequences will show degrees of variation dependent on the degree of relatedness of those individuals. Once rRNA sequences have been determined, computer analysis allows determination of sequences and will detect those that are unique to any particular organism (signature sequences).

Restriction analysis. Chromosomal DNA preparations digested by restriction enzymes will be cleaved into smaller fragments. These fragments can be separated by agarose gel electrophoresis and it is then possible to map the restriction sites for the enzymes used. Restriction sites will be similar for organisms of the same species and the lengths of the fragments formed can be compared. This is known as restriction fragment length polymorphism (RFLP) analysis.

8.7 Depositing in culture collections

Once an interesting strain or new species of microbe has been isolated, identified and grown in culture, it becomes very important to maintain and store subcultures in an appropriate way (Chapter 9). Additionally, it may be important to consider depositing the culture in a culture collection where it will be maintained in the long term. A safe-deposit in a collection provides a professional system for maintaining important cultures but confidentiality and ownership of the deposit is retained by the depositor. Deposit in an open collection allows the culture to be made available for the use of other research workers.

The policies of a particular collection for acquiring cultures vary depending on the scientific role of that collection. Some store only fungi, some serve bacterial systematics and others have a more general purpose. The decision of a culture collection to acquire a particular culture may be influenced by the ease with which that culture could be maintained within the normal routines of the collection and also by the remit of the funding body. Individual collections should be consulted concerning their particular policy [9].

A newly isolated microbe must have a detailed description and the scientific name must be indicated in order for the information to be published (see Section 8.1). A culture of the organism should be preserved as a dried specimen or a slide preparation. In addition to that, a live culture may be deposited in a collection. Acquisition of a culture is a long-term commitment by the collection, and as full documentation as possible is required. Information about the properties of the isolate, including copies of published scientific papers, where the organism has been cited, will be retained together for reference. If the organism produces unusual secondary products on particular media, details are required. Some collections issue specific forms to be filled in at the time of deposition.

References

1. Skerman, V.B.D., McGowan, V. and Sneath, P.H.A. (1989) *Approved Lists of Bacterial Names* (amended edn). American Society for Microbiology, Washington, DC.
2. Moore, W.E.C. and Moore, L.V.H. (1989) index published in *International Journal of Systematic Bacteriology*. American Society for Microbiology, Washington, DC.
3. Holt, J.G., Kreig, N.R., Sneath, P.H.A., Staley, J.T. and Williams, S.T. (eds) (1994) *Bergey's Manual of Determinative Bacteriology* (9th edn). Williams and Wilkins, Baltimore, MD.
4. Hawksworth, D.L., Sutton, B.C. and Ainsworth, G.C. (eds) (1983) *Ainsworth and Bisby's Dictionary of the Fungi* (7th edn). Commonwealth Mycological Institute, Kew, Surrey.
5. Rawlins, D.J. (1992) *Light Microscopy*, Introduction to Biotechniques Series. BIOS Scientific Publishers, Oxford.
6. Gerhardt, P., Murray, R.G.E., Costilow, R.N., Nester, E.W., Wood, W.A., Kreig, N.R. and Phillips, G.B. (eds) (1981) *Manual of Methods for General Bacteriology*. American Society for Microbiology, Washington, DC.
7. Gerhardt, P., Murray, R.G.E., Wood, W.A. and Kreig, N.R. (eds) (1994) *Methods for General Bacteriology*. American Society for Microbiology, Washington, DC.
8. Priest, F. and Austin, B. (1993) *Modern Bacterial Taxonomy* (2nd edn). Chapman & Hall, London.
9. Hawksworth, D.L. and Kirsop, B.E. (eds) (1988) *Living Resources for Biotechnology: Filamentous Fungi*. Cambridge University Press, Cambridge.

9 Cultures: their Preservation and how to Obtain them from Other Sources

9.1 Introduction

If you are maintaining a particular organism for your own particular needs, it is self-evident that you maintain it alive, but it is also necessary to maintain it in a physiological state as near as possible to that of the original isolate. It may be necessary to have the organism maintained in culture long after you have finished working with it, in the eventuality that someone else may wish to work experimentally with the same isolate or compare it with other organisms for taxonomic purposes. Thus, in terms of maintaining a culture there are short- and long-term methods of preservation. Many of the methods are as applicable to fungi as to bacteria, so much of what is said below is relevant to both groups of organisms. Where relevant, attention will be drawn to the special requirements of particular groups of organisms.

This chapter also indicates how you may obtain information about collections from which cultures can be ordered for your own use. There is also advice about the procedures for dispatching and importing cultures.

9.2 Subculturing for experimental purposes

Those who are routinely using the same organism(s) for experimental work must maintain their organism(s) by subculturing, namely the periodic transfer to fresh media. It is frequently suggested that minimal medium be used for subculturing, in order to prolong the time between transfers. But if the organism is being used experimentally, there are strong arguments, in terms of physiological constancy, for using a medium, the composition of which forms the basis of the media used in experimental work. While the use of such a medium might mean more frequent transfers, there will be little change in metabolic capability of the organism between one experiment and another.

Most frequently, cultures are maintained on nutrient agar slopes in tubes with screw caps with rubber inserts to reduce water loss. But those working with filamentous fungi often use Petri dishes sealed with laboratory sealing tape, for example Parafilm. Petri dish cultures allow much easier access for subculturing filamentous fungi and, furthermore, it is easier to ensure that one is sampling from the colony margin.

Some points to remember when involved in a regime of subculturing are listed as follows.

- It is preferable to do it to an established routine which is known to ensure effective establishment of new cultures. This means that subculturing should be carried out at set intervals and the cultures maintained at the same temperature.
- Cultures should be checked routinely for any changes in phenotype. In this respect, select several colonies of bacteria or yeasts for making a subculture; selection of only a single colony leads to a greater chance of selecting mutants. In the case of filamentous fungi, it is sensible to have several colonies (in separate tubes/Petri dishes) from which subcultures can be made. Discard any showing visible signs of being different.
- Do not use all your cultures when subculturing. If the new cultures become contaminated, you can start again with one of the cultures that has been set aside. As a further precaution, it is worthwhile to have two or more cultures of the particular organism(s) being used in your experiments under conditions that minimize the need to subculture (see below). Such cultures insure against complete loss of those cultures which are supporting daily experimental work.
- Cultures of the organism should be placed in different stores in the laboratory, to protect against loss by accidental change in storage conditions, for example unforeseen temperature change.

9.3 Subculturing a small collection of micro-organisms

If you have a small collection of cultures (say 25–50) which is used infrequently, then you should use procedures to reduce the need for subculturing. The most frequently used procedures are listed as follows.

- *Storage in a refrigerator.* The time prior to subculturing (compared to storage in an incubator at 20°C) can be approximately doubled by storage in this way.
- *Storage under oil.* This procedure with slant cultures is of long standing. The culture is covered with medicinal liquid paraffin (mineral oil) which should be sterilized previously by heating in an oven at 170°C for 1–2 h; it should not be autoclaved. The culture is allowed to grow until about half the surface of the agar is covered, then the oil is added to a depth of around 1 cm. All the agar must be covered as evaporation from an exposed portion can dry it out.
- *Use of soil.* Spore-forming bacteria and fungi can often be kept in culture in soil. The soil is first air-dried by spreading thinly on a large, flat surface, then put into screw-topped vials and autoclaved on two successive days. The vials can then be inoculated with 1 ml of a suspension of the organism in sterile distilled water and incubated for a period (1–2 weeks) at 20°C for growth to occur through the soil. These cultures are then stored in the refrigerator with the caps slightly loose. The organism is recovered by sprinkling a small quantity of soil on to a solid medium that gives optimum growth and incubating at the temperature for such growth.

Other procedures are essentially suitable only for particular groups of organisms. You are recommended to consult specialist texts for the necessary information.

9.4 Long-term culture methods

Reliable long-term methods are required by culture collections. If you have an organism which you have brought into culture, investigated and the results of the investigation are in the published literature, then you should consider depositing a culture in such a collection. This should allow your investigations to be repeated and built upon by someone else at a later date. The procedures used in collections to maintain cultures are designed to maintain a culture in a stable condition. The focus is, of course, on genetic stability. Two methods are used: *freeze-drying* and *freezing*.

Freeze-drying (lyophilization) is a very successful method which involves the removal of water from a microbial culture by sublimation under reduced pressure. Success depends on the prevention of damage by the process of freezing and as a result of oxidation. The organism should be suspended in a cryoprotective agent. A range of such agents has been used, such as skimmed milk, glucose-horse serum, inositol-serum, inositol broth, dextran and sucrose. Cultures benefit from slow cooling, for example 1°C min^{-1} and should be kept below −15°C until the operation is completed. Completion does not mean complete dryness; around 1% residual water must be present. Cultures are frozen in vials which can be sealed such that the vacuum is maintained. Ampoules sealed in this way can be distributed easily by post, since with the appropriate vial design and packaging there is minimal risk of the culture being a danger or it being contaminated. Nevertheless, there can be problems with the stability of the cultures. A range of fungi, particularly hyphal basidiomycetes, are sensitive to freeze-drying. With a viable culture, rehydration may take several hours but is usually complete within 30 min after addition of the appropriate liquid nutrient medium (see Section 9.6).

With regard to the second method, *freezing*, if micro-organisms are frozen, no further metabolism occurs when all the internal water is frozen. However, ice recrystallization can occur above −130°C and this can damage the frozen organism. Therefore the most effective storage of micro-organisms is at the temperature of liquid nitrogen, namely −196°C. Nevertheless, storage at −70°C can be effective for many micro-organisms. The rates of cooling should be slow, for example 1°C min^{-1}; in contrast to recovery, when a fast rate of warming (ca. 200°C min^{-1} is recommended. The suspending medium is also important and cryoprotectants are usually used. These are nontoxic, penetrate the organism rapidly and reduce the possibility of ice crystal formation. Compounds used essentially act as antifreeze agents, such that the freezing point of the protoplasm is very considerably lowered. Also, the compounds replace the water in the organism without having detrimental effects on protein conformation and membrane stability. Effective cryoprotectants are often those compounds that allow micro-organisms to withstand conditions of low water availability; examples are glycerol, arabitol, proline and trehalose. Dimethylsulfoxide is also a very good cryoprotectant; it and glycerol are the compounds most used in the storage of micro-organisms by freezing. Smith [1] provides a very good account of how freezing and thawing affect fungi, and about the general principles that underlie cryoprotection.

Table 9.1 provides a comparison of the various methods described in the text for the preservation of micro-organisms.

TABLE 9.1: *Comparison of methods of preserving micro-organisms [2]*

Method of preservation	Cost		Shelf-life	Genetic stability
	Material	**Labor**	**Shelf-life**	
Serial transfer on agar				
Storage at room temperature	Low	High	1–6 months	Variable
Storage in the refrigerator	Medium[a]	High	6–12 months	Variable
Storage under oil	Low	Low/medium	1–32 years	Poor
Storage in water	Low	Low/medium	2–5 years	Moderate
Storage in the deep freeze	Medium[a]	Low/medium	4–5 years	Moderate
Drying				
Soil	Low	Medium	5–20 years	Moderate to low
Freeze-drying	High	Initially medium[b]	4–40 years	Good
Freezing				
Liquid nitrogen storage	High	Low	Infinite	Good

[a] Refrigerator or deep-freeze costs included.
[b] Initial processing is costly, depending on the method; subsequent storage costs are negligible.

9.5 Obtaining and transporting cultures

In Appendix B we give addresses of major culture collections. For others, we suggest you consult the *Living Resources for Biotechnology* series listed in Appendix C.

If you are sending a culture to another laboratory for the first time, it is sensible to obtain expert advice about procedures and also about the appropriate container to use for the culture and the packaging needed to give full protection. You must also seek advice from your national postal service about the conditions necessary for the acceptance of 'infectious perishable biological substances'. The despatch of any perishable biological material in the overseas post by yourself will be restricted to those countries whose postal administrations will accept such items.

You need to be aware that you will almost certainly need a license to import plant pathogens. The organization providing the license will depend on the country in which you work. For instance, in the UK, such a license is issued by the Ministry of Agriculture, Fisheries and

Food and covers the ability of the institute receiving pathogenic fungi to handle the fungus and infected plants in such a way that the ability of the fungus to be dispersed elsewhere is effectively contained. Remember, it is the natural ability of the fungus which is at issue not the use to which it is put in the laboratory. Thus, it is almost certain that even if you require the pathogenic fungus for reasons other than an interest in plant pathology, a license will still be required.

9.6 Dealing with a culture when it arrives

If the culture is in a screw-top tube or similar container, the procedure will be exactly the same as for subculturing from such a container. If a culture is from an established culture collection, it is likely that it will have been freeze-dried. The procedure for dealing with such a culture is outlined as follows.

- Freeze-dried cultures are normally contained in sealed ampoules (glass tube). The culture is usually at the rounded end, while the opposite end will have been sealed and there will be a cotton wool plug near the center of the tube.

- Using aseptic techniques throughout, the ampoule should be scored with a metal file around the circumference at the middle of the cotton wool plug. The ampoule may then be cracked open (using ampoule breakers for safety) or alternatively a hot glass rod may be placed close to the score mark to crack the glass.

- Once entry has been achieved, the culture will need to be rehydrated as soon as possible. Carefully remove and retain the cotton wool plug with your little finger, and add a few drops (ca. 0.5–1.0 ml) of sterile distilled water to the culture. Replace the plug and allow the tube to rest for about 30 min in a slightly inclined position (to prevent the liquid from coming in contact with the cotton wool). After rehydration, the contents of the ampoule should be well mixed in a suspension, all of which should be poured on to a suitable growth medium. The culture should be incubated under those conditions known to favor good growth.

- Initially, growth of the culture will be slow. Normal rates should be restored at the next subculture. At this stage, replicate subcultures should be grown for storage and the growth characteristics of the organism should be properly documented for further reference.

References

1. Smith, D. (1993) in *Stress Tolerance of Fungi* (D.H. Jennings, ed.). Marcel Dekker, New York.
2. Smith, D. (1988) in *Living Resources for Biotechnology: Filamentous Fungi* (D. L. Hawkesworth and B.E. Kirsop, eds). Cambridge University Press, Cambridge, pp. 75–89.

Appendix A

Glossary

Acidophile: a micro-organism that thrives in relatively acid media.

Aerobe: a micro-organism that requires oxygen for growth.

Agar: a complex polysaccharide derived from the cell walls of seaweeds which forms a gel when a hot dilute solution (1–2% v/v) is allowed to cool.

Alkalophile: a micro-organism that can be considered to be unable to grow at pH 7 or less, has optimal growth around pH 9 and is capable of growth above pH 10.

Anaerobe: a micro-organism that does not require oxygen for growth.

Anamorph: the form of a particular fungus that only exhibits asexual reproduction (*see* Teleomorph).

Aseptic: free from the presence of (micro-)organisms.

Autoclaving: sterilization by steam under a pressure greater than atmospheric (and therefore also a temperature higher than 100°C).

Autotroph: a microbe whose nutrients are completely inorganic.

Axenic culture: the presence of only one (micro-)organism.

Biotrophy: invasion of living cells by a microbe and obtaining nutrients from them without immediately killing them.

Broth: undefined (q.v.) liquid medium.

Chemolithotroph: a micro-organism that obtains its energy for growth from an inorganic compound(s).

Chemo-organotroph: a micro-organism that obtains its energy for growth from an organic compound(s).

Coccus (plural **cocci**): a spherical bacterial cell.

Coenocyte: a cell or hyphal compartment containing more than one nucleus.

Colony: an aggregation of microbial cells or, in the case of filamentous fungi, the mycelium produced from a single spore or mycelial fragment.

Conidium (plural **conidia**): a specialized fungal spore which is produced at, and is deciduous from, the end of an essentially vegetative hypha.

Containment: the procedures and facilities required for the safe handling of micro-organisms. The nature of the containment depends on the hazardous nature of the micro-organisms being handled.

Defined medium: a medium in which all the constituents and their individual concentrations are known.

Denitrification: the breakdown of nitrogen compounds to nitrogen gas.

Diagnostic tests: simple and easily produced tests that allow the user of them to characterize a particular microbe in terms of the ability to utilize an organic compound for growth, grow in the absence of oxygen, etc.

Dimorphism: the ability of a fungus to exist in a cellular or filamentous form.

Endospore: a spore produced within a bacterial cell.

Erlenmeyer flask: conical flask.

Eukaryotes: organisms with a nuclear membrane and 'linear' chromosomes (*see Table 1.2*).

Facultative anaerobe: a micro-organism that grows well in the absence of oxygen but suffers no detrimental effects on exposure to air.

Flagellae (singular **flagellum**): filaments extending from cells to which they give motility. The flagellae of prokaryotes differ fundamentally in stucture from those of eukaryotes.

Growth factor: a compound which, in small amounts, is necessary or stimulatory for growth but does not serve as an energy source.

Halophiles: organisms that can complete their life cycle in the presence of a relatively high concentration (ca. 5000 mM) of NaCl.

Hanging drop culture: a preparation made using a microscope slide with a small depression in it; a drop of culture is suspended from a coverslip over the depression and then observed microscopically.

Heterocyst: a cell (part of a cyanobacterial filament) which is specialized for fixing atmospheric nitrogen.

Heterotroph: a microbe requiring organic carbon as a source of energy.

Hypha (plural **hyphae**): a fungal filament containing cytoplasm, with its full complement of organelles, which may or may not be divided by septa, and which extends at its end (tip/hyphal apex) and is capable of branching.

Hyphal growth unit: the unit of protoplasm in a filamentous fungus that doubles in amount over time like a unicell but manifests itself as a new hyphal apex.

Inoculum: the source of a new/fresh culture, which may be in the form of a suspension of cells/spores/hyphal fragments or pieces of mycelium.

Laminar flow cabinet (sterile hood): a cabinet allowing aseptic handling of microbes because of the flow through it of sterile air.

Major element: one that is required at a concentration around 10^{-2} M or higher for growth of micro-organisms.

Maximum temperature: the temperature above which an organism will not grow.

Membrane filters: sieves for solutes and/or supended material, composed of cellulose acetate.

Methanogens: micro-organisms producing methane.

Microaerophile: a micro-organism that grows best at oxygen levels lower than in the air.

Minimum temperature: the temperature below which an organism will not grow.

Minor element: one that is required at a concentration around 10^{-5}–10^{-3} M for growth of micro-organisms.

Mutualism: *see* Symbiosis

Mycelium (plural **mycelia**): a colony produced by the extension of a hypha, originating from a spore or hyphal fragment, and the branches that arise subsequently.

Mycobiont: the fungal partner in a lichen (*see* Phycobiont).

Necrotrophy: invasion and killing of the tissues of a multicellular organism, and feeding thereon, by a microbe.

Niche: a part of the environment, both spatially and temporally, that allows a particular organism to grow and reproduce successfully.

Nucleopore filters: polycarbonate filters that have precise pore sizes.

Optimum temperature: the temperature at which there is the most rapid growth of the organism.

Parasexual cycle: genetic recombination taking place within the mitotic cycle; it occurs in some fungi that have lost their sexual stage.

Parasite: an organism that lives in association with, and benefits from, another dissimilarly named organism.

Pasteur pipette: a glass tube drawn out such that it has a very narrow diameter for most of its length; liquid is drawn up and expelled by pressure on a rubber teat at the much wider end.

Pathogen: an organism that brings about either observable damage to or reduced growth of another dissimilarly named organism with which it is directly associated and from which it benefits.

Pellet: an essentially spherical mycelium produced by growth of a filament of fungus in liquid culture.

Peripheral growth zone: the extent of the mycelium that is required to support maximal extension of those hyphae at the margin of a filamentous fungal colony.

Petri dish: a round, clear plastic or, less frequently, glass dish with a lid that fits well and overlaps the base.

Petri plate: a Petri dish containing a culture on nutrient agar.

Phototroph: a micro-organism that derives its energy for growth from light.

Phycobiont: the algal or cyanobacterial partner in a lichen (*see* Mycobiont).

Prokaryotes: organisms possessing a chromosome (*see Table 1.2*) but without a nuclear membrane.

Saprotrophy: growth of a microbe on nonliving organic materials.

Selective media: media designed to support the growth of a particular micro-organism(s) at the expense of others that might be present in the inoculum.

Septum (plural **septa**): a cross wall, which may or may not contain a pore or pores, in a cell or hypha.

Shake cultures: liquid cultures agitated on a shaker, usually with a rotary action.

Slope culture: a culture growing on agar medium set at an angle in a glass tube.

Spore: a propagule that has been specially produced for propagating/reproducing an organism; it may be produced either asexually or sexually.

Stab culture: one produced by introduction of the organism on a long metal wire into a tube of agar; the position of growth down the stab indicates the organism's requirement of oxygen, or lack of it, for growth.

Sterile hood: *see* Laminar flow cabinet.

Symbiosis: the living of one organism in association with another dissimilarly named organism, to the mutual benefit of both organisms. Such an association has also been described as one of **mutualism**.

Teleomorph: the form of a particular fungus that exhibits sexual reproduction (*see* Anamorph).

Trace element: one that is required at a concentration much lower than that required for a minor element (qv).

Undefined medium: a medium based usually on natural complex materials, such that either all the potential individual nutrients cannot be known or, if they are known, their concentration is uncertain.

Vibrio: a curved bacterial cell.

Vitamin: a compound that is required in small amounts for growth but does not function as a structural material and is not used as an energy source.

Washout: removal of cells from continuous culture by inflow of medium at a rate faster than the ability of the organism to divide to maintain the concentration of cells.

Water activity: this is the vapor pressure above the solution, medium or organism relative to the standard state (*see* Water potential).

Water potential: a measure of the force driving water into a cell or hypha. Thermodynamically, it is the chemical potential divided by the partial molar volume. Water potential is logarithmically related to water activity.

Appendix B

Culture collections

Address	Cultures maintained
American Type Culture Collection (ATCC) 12301 Parklawn Drive Rockville Maryland, 20852 USA	Algae, bacteria, fungi, protozoa, yeasts
Centraalbureau voor Schimmelcultures (CBS) Oosteraat 1 Baarn The Netherlands	Actinomycetes, bacteria, fungi, yeasts
Culture Collection of Algae and Protozoa (CCAP) Freshwater Biological Association Windermere Laboratory The Ferry House Far Sawrey, Ambleside Cumbria LA22 0LP UK	Freshwater and terrestrial algae, protozoa and marine algae
Deutsche Sammlung von Mikroorganismen (DSM) Grisebachstrasse 8 3400 Gottingen Germany	Actinomycetes, bacteria, fungi, yeasts
International Mycological Institute (IMI) Bakeham Lane Egham Surrey TW20 9TY UK	Fungi, yeasts
National Collection of Industrial Bacteria (NCIB) c/o The National Collections of Industrial and Marine Bacteria Ltd Torry Research Station P.O. Box 31 135 Abbey Road Aberdeen AB9 8DG UK	Bacteria

References

For details of other collections see:

Hawksworth, D.L. and Kirsop, B.E. (eds) (1988) *Living Resources for Biotechnology: Filamentous Fungi.* Cambridge University Press, Cambridge.

Holt, J.G., Kreig, N.R., Sneath, P.H.A., Staley, J.T. and Williams, S.T. (eds) (1994) *Bergey's Manual of Determinative Bacteriology* (9th edn). Williams and Wilkins, Baltimore, Maryland. (All volumes contain a list of collections of prokaryotes.)

Appendix C

Further reading

Collins, C.H., Lyne, P.M. and Grange, J.M. (1979) *Collin and Lyne's Microbiological Methods* (4th edn). Butterworths, London.

Fleming, D.O., Richardson, J.H., Tulis, J.J. and Vesley, D. (eds) (1995) *Laboratory Safety: Principles and Practices*. American Society for Microbiology, Washington, DC.

Gerhardt, P., Murray, R.E.F., Costilow, R.N., Nester, E.W., Wood, W.W., Kreig, N.R. and Phillips, G.B. (eds) (1981) *Manual of Methods for General Bacteriology*. American Society for Microbiology, Washington DC.

Hawksworth, D.L. and Kirsop, B.E. (eds) (1988) *Living Resources for Biotechnology: Filamentous Fungi*. Cambridge University Press, Cambridge.

Hill, L. and Kirsop, B.E. (eds) (1991) *Living Resources for Biotechnology: Bacteria*. Cambridge University Press, Cambridge.

Kirsop, B.E. and Doyle, A. (eds) (1991) *Maintenance of Microorganisms and Cultured Cells*. Academic Press, London.

Kirsop, B.E. and Kurtzman, C.P. (eds) (1988) *Living Resources for Biotechnology: Yeasts*. Cambridge University Press, Cambridge.

Logan, N.A. (1994) *Bacterial Systematics*. Blackwell Scientific Publications, Oxford.

Meynell, G.G. and Meynell, E. (1970) *Theory and Practice in Experimental Bacteriology*. Cambridge University Press, Cambridge.

Norris, J.R. and Ribbons, D.W. (eds) (1969) *Methods in Microbiology*. Academic Press, London, Vol. 1 *et seq*. (This series, still being published and sometimes with other editors, is a mine of information for the microbiologist working in the laboratory; the first four volumes are particularly relevant to anyone starting laboratory work in microbiology.)

Pirt, S.J. (1975) *Principles of Microbe and Cell Cultivation*. Blackwell Scientific Publications, Oxford.

Index